EMS PRESS

Zurich Lectures in Advanced Mathematics

Mathematics in Zurich has a long and distinguished tradition, in which the writing of lecture notes volumes and research monographs plays a prominent part. The *Zurich Lectures in Advanced Mathematics* series aims to make some of these publications better known to a wider audience. The series has three main constituents: lecture notes on advanced topics given by internationally renowned experts, in particular lecture notes of "Nachdiplomvorlesungen", organized jointly by the Department of Mathematics and the Institute for Research in Mathematics (FIM) at ETH, graduate text books designed for the joint graduate program in Mathematics of the ETH and the University of Zürich, as well as contributions from researchers in residence. Moderately priced, concise and lively in style, the volumes of this series will appeal to researchers and students alike, who seek an informed introduction to important areas of current research.

Previously published in this series:

Y. B. Pesin, *Lectures on partial hyperbolicity and stable ergodicity*
S.-Y. A. Chang, *Non-linear Elliptic Equations in Conformal Geometry*
S. B. Kuksin, *Randomly forced nonlinear PDEs and statistical hydrodynamics in 2 space dimensions*
P. Etingof, *Calogero–Moser systems and representation theory*
G. Balkema, P. Embrechts, *High Risk Scenarios and Extremes – A geometric approach*
D. Christodoulou, *Mathematical Problems of General Relativity I*
C. De Lellis, *Rectifiable Sets, Densities and Tangent Measures*
P. Seidel, *Fukaya Categories and Picard–Lefschetz Theory*
A. H. W. Schmitt, *Geometric Invariant Theory and Decorated Principal Bundles*
M. Farber, *Invitation to Topological Robotics*
A. Barvinok, *Integer Points in Polyhedra*
C. Lubich, *From Quantum to Classical Molecular Dynamics: Reduced Models and Numerical Analysis*
S. Onn, *Nonlinear Discrete Optimization – An Algorithmic Theory*
K. Nakanishi, W. Schlag, *Invariant Manifolds and Dispersive Hamiltonian Evolution Equations*
E. Faou, *Geometric Numerical Integration and Schrödinger Equations*
A.-S. Sznitman, *Topics in Occupation Times and Gaussian Free Fields*
F. Labourie, *Lectures on Representations of Surface Groups*
I. Gallagher, L. Saint-Raymond, B. Texier, *From Newton to Boltzmann: Hard Spheres and Short-range Potentials*
R. J. Marsh, *Lecture Notes on Cluster Algebras*
E. Hebey, *Compactness and Stability for Nonlinear Elliptic Equations*
S. Serfaty, *Coulomb Gases and Ginzburg–Landau Vortices*
A. Figalli, *The Monge–Ampère Equation and Its Applications*
W. Schachermayer, *Asymptotic Theory of Transaction Costs*
A. Thomas, *Geometric and Topological Aspects of Coxeter Groups and Buildings*
T. Fisher, B. Hasseblatt, *Hyperbolic Flows*
S. Baader, *Geometry and Topology of Surfaces*
N. Anantharaman, *Quantum Ergodicity and Delocalization of Schrödinger Eigenfunctions*
X. Fernández-Real, X. Ros-Oton, *Regularity Theory for Elliptic PDE*
A. Polishchuk, *A_∞-Structures and Moduli Spaces*
R. Nickl, *Bayesian Non-Linear Statistical Inverse Problems*

Published with the support of the Huber-Kudlich-Stiftung, Zürich

Andrea Malchiodi

Prescribing Scalar Curvature in Conformal Geometry

EMS PRESS

Author

Andrea Malchiodi
Scuola Normale Superiore
Piazza dei Cavalieri, 7
56126 Pisa, Italy

Email: andrea.malchiodi@sns.it

2020 Mathematics Subject Classification: Primary 58J05; Secondary 35B33, 35J60, 35J20, 58E05, 53C21, 53-02

Keywords: prescribed scalar curvature, elliptic equations of critical type, blow-up analysis, topological methods

ISBN 978-3-98547-052-5, eISBN 978-3-98547-552-0, DOI 10.4171/ZLAM/31

Bibliographic information published by the Deutsche Nationalbibliothek
The Deutsche Nationalbibliothek lists this publication in the Deutsche Nationalbibliografie; detailed bibliographic data are available on the Internet at http://dnb.dnb.de.

Published by EMS Press, an imprint of the

European Mathematical Society – EMS – Publishing House GmbH
Institut für Mathematik
Technische Universität Berlin
Straße des 17. Juni 136
10623 Berlin, Germany

https://ems.press

Typeset by Alison Durham, Manchester, UK
Printed in Germany
♾ Printed on acid free paper

To Elisa and Bianca

Preface

These notes originate from a Nachdiplomvorlesung held by the author at ETH during the spring term of 2021. The main focus of the course was the classical problem initiated by Kazdan and Warner in the 1970s, about prescribing a given function on a closed manifold M as the scalar curvature of some Riemannian metric. An interesting case of the problem is when one looks for such a metric within a given conformal class, which makes the problem equivalent to solving an elliptic PDE involving the critical Sobolev exponent. This feature results in a loss of compactness, which is typical of scaling-invariant problems. The issue can be tackled by combining several techniques, like exploiting the variational structure of the problem, the use of blow-up analysis, and asymptotic expansions, as well as employing topological tools such as min-max, Morse, or Leray–Schauder degree theory. The aim of the course was to give a general overview of the problem, to present some key results and techniques, and to cover some of the progress on this topic, ranging from the original contributions by Kazdan–Warner, to the most recent results on the subject.

The notes are structured as follows. In the introduction we present the two main versions of the Kazdan–Warner problem, illustrate the variational strategy and the blow-up analysis technique, and state the main results we will focus on. Chapter 2 collects some preliminary material: basic notions in differential geometry, useful functional spaces on compact manifolds, and some elliptic regularity theory, both at linear and non-linear levels. In Chapter 3 we will present the covariance property and some spectral features of the *conformal Laplacian* and the *Yamabe quotient*, which characterize conformal classes of metrics. An exact expression for the extremal value of the quotient will be given for the round sphere, and the resolution of the Yamabe conjecture will be discussed. We will then show the solvability of the Kazdan–Warner problem in its most general form, without any restriction to a conformal class.

We will then turn to the conformally constrained Kazdan–Warner problem in dimension $n \geq 3$, which will indeed be our main focus. As we remarked above, this version of the problem will amount to solving an elliptic equation with a non-linearity of *critical type* on the underlying manifold. In Chapter 4 we will introduce the variational setting for treating the conformal Kazdan–Warner problem. This will lead to full solvability in the case of non-positive Yamabe class by a global minimization procedure. We will show that this approach is not possible in positive Yamabe class, but that in some symmetric situations this property might be recovered, illustrating a result by Escobar and Schoen.

In Chapter 5 a classification theorem for the Yamabe equation on $\mathbb{R}^n$ will be proved, using a variant of the classical moving plane method. We will then turn to sub-critical equations, for which a Liouville-type theorem will also be shown, useful to prove a priori estimates. These results will be crucial to study the phenomenon of loss of compactness, typical of geometric non-linear equations. In Chapter 6 we will perform a blow-up analysis on solutions of sub-critical equations with large L^∞-norm on compact manifolds: it will be shown that in dimensions 3 and 4, blow-ups are *isolated simple*, and that this condition persists in higher dimensions for sequences of solutions with zero weak limit and uniformly bounded Sobolev norm. In Chapter 7 a converse result will be proven, and solutions developing isolated simple blow-ups for sub-critical approximations of the prescribed curvature problem will be constructed using a finite-dimensional reduction. These techniques can also be used in general dimension to perform blow-up analysis of particular classes of solutions.

In the eighth chapter we then present an application of the previous two, combined with some topological tools. In low dimension we present a general existence theory for the scalar curvature prescription problem, and in arbitrary dimension also under suitable pinching conditions on the curvature function. Chapter 9 is then devoted to some results about non-existence, showing the sharpness of the assumptions in the theorems presented in Chapter 8. Such results are built over a classical obstruction due to Kazdan and Warner for the sphere, which is extended with the aid of blow-up analysis. Finally, in Chapter 10 we present some related research directions and open questions.

Many contributions from the literature on this topic, which is vast, cannot unfortunately be covered in these notes. We hope anyway to give an opportunity for the interested reader to get a general view of the subject and to create occasions for further exploration. Some of the most technical parts of some proofs are omitted for brevity, but a precise account of the sources with full details will be given.

I am deeply grateful to Alessandro Carlotto and Michael Struwe for having suggested me as a lecturer in the Nachdiplomvorlesung series, and to Alessio Figalli and Andrea Waldburger for making the visit possible and pleasant, despite the difficulties generated by the pandemic period. Special praise goes to Ali Hyder for his precious help in the editing of these notes. I would also like to thank Alessandro Carlotto, Matt Gursky, and Martin Mayer for their valuable comments on a first draft. Finally, I would like to thank all the participants who attended the lectures for their useful observations and for creating a nice atmosphere on the few occasions of gathering that were permitted by the circumstances.

Contents

Chapter 1

Introduction

A classical question in differential geometry is the following one, posed in [140, 144]:

(KW-1) *Given a manifold* M *and a function* $K: M \to \mathbb{R}$, *find a Riemannian metric* $\hat{g}$ *on* M *such that the scalar curvature* $R_{\hat{g}}$ *of* $\hat{g}$ *coincides with* K.

In general, the answer to the above problem is negative. For example, in dimension $n = 2$, the scalar curvature is twice the Gaussian curvature. Therefore, if M is compact without boundary, by the Gauss–Bonnet theorem,

$$\int_M R_g \, dV_g = 4\pi\chi(M).$$

In particular, if there is a solution to the prescribed curvature problem, somewhere K has to possess the sign of $\chi(M)$.

Another version of the above problem is the restriction to a fixed conformal class:

(KW-2) *Given a Riemannian metric* g *on* M, *try to solve* (KW-1) *with a metric* $\tilde{g}$ *in the* conformal class *of* g, *i.e.* $\tilde{g}$ *is of the form* $\tilde{g}(x) = \Lambda(x)g(x)$, *where* Λ *is a smooth positive function on* M.

Main focus of these notes. We will assume that (M^n, g) is *closed*, meaning – unless specified – *compact without boundary and of dimension* $n \geq 3$. We mainly deal with the conformal version (KW-2) of the above problem. In fact, as we will see in Chapter 3, (KW-1) is solvable under rather mild requirements on the prescribed curvature function K.

There are both conceptual and technical differences for the case $n = 2$. We will make several comments here and there in this regard.

Remark 1.1. We point out the following facts:

(i) There exists a version of the problem between (KW-1) and (KW-2): this consists in prescribing the scalar curvature combining conformal changes and pull-backs of the metric via diffeomorphisms on M. In the terminology of Kazdan–Warner, such changes of metric are referred to as *conformal equivalences*, but *not* in our notation!

(ii) The following are interesting particular cases of (KW-2) (conformal case):

- Yamabe's problem [218]: on (M^n, g) closed with $n \geq 3$, find a conformal metric with constant scalar curvature. By [15, 193, 211] this is always solvable. For $n = 2$ this is the classical *uniformization problem*, initiated by Klein, Koebe, and Poincaré.

- Nirenberg's problem: on $(M^n, g) = (\mathbb{S}^n, g_{\mathbb{S}^n})$, prescribe conformally a non-constant function. This is a well-studied problem, originally posed for $n = 2$, and we will give an account in the notes of several results on this research line.

The books [16,65,97,130,131,209] are particularly useful references for the purposes of these notes, and contain detailed presentations of some of the topics covered here.

1.1 Problem (KW-1)

Even though we will mainly focus on (KW-2), we will illustrate the solvability of the first version of the Kazdan–Warner problem since its proof is rather self-contained, apart from the fact that it requires the resolution of Yamabe's problem. This will not be discussed in detail here, even though we will present the main ingredients needed to prove it. We begin by introducing the following definition.

Definition 1.2. A closed manifold M of dimension $n \geq 3$ is said to be of *type I* if it admits a metric of non-negative and non-identically zero scalar curvature. It is said to be of *type II* if it admits a metric of non-negative scalar curvature, but any such metric has scalar curvature identically equal to zero. The manifold M is said to be of *type III* if the scalar curvature of any metric on M has to be negative somewhere.

An important fact to note is that if M^n is closed and of dimension $n \geq 3$, then by a result in [13] there always exists a metric $\bar{g}$ with $R_{\bar{g}} < 0$. Indeed, there even exists $\bar{g}$ with $\mathrm{Ric}_{\bar{g}} < 0$ [110, 160]. However, not all manifolds carry a metric with positive scalar curvature: understanding when it happens is a hard problem, with important contributions such as [116, 197, 206]; see e.g. the recent survey [50].

The flexibility in the choice of metric for problem (KW-1) yields the following general result, by which it is possible to prescribe a large class of functions.

Theorem A ([141–143]). *Suppose M^n is closed, with $n \geq 3$. If M is of type I, then* (KW-1) *is solvable for every $K \in C^\infty(M)$. If M is of type II, then* (KW-1) *is solvable for every $K \in C^\infty(M)$, which is either negative somewhere or identically zero. If M is of type III, then* (KW-1) *is solvable for every $K \in C^\infty(M)$ that is negative somewhere.*

Concerning the first statement of the latter theorem, Kazdan and Warner assumed the existence of metrics with positive and constant scalar curvature. Indeed, at the time this was not known to hold in general, but was later proved in [15, 193, 211] for every conformal class, and hence Theorem A can be stated without any extra assumptions.

Remark 1.3. We note the following:

(i) For $n = 2$ and M closed, (KW-1) is solvable provided somewhere K attains the sign of $\chi(M)$.

(ii) If M^n, $n \geq 3$ is non-compact and diffeomorphic to an open set U of a compact manifold $\bar{M}$, then (KW-1) is solvable for every $K \in C^\infty(M)$.

The proof of Theorem A relies on tools that are of functional-analytic nature, and exploits the invertibility of the operator that assigns to each conformal factor the scalar curvature of the metric it induces. The idea is then, starting from a metric with constant scalar curvature, to get any other function close to that constant in L^p to it by composing with a proper diffeomorphism on M. In positive curvature, this argument needs a special spectral property of the linearized equation for the scalar curvature, guaranteed by the above-mentioned results by Trudinger, Aubin, and Schoen.

1.2 Problem (KW-2)

Recall that for this version of the problem one tries to prescribe a given function on M as the scalar curvature via conformal deformations of a background metric g. As will be shown in Chapter 2, for $\tilde{g}(x) = u(x)^{\frac{4}{n-2}} g(x)$, $n \geq 3$, the scalar curvature R_g transforms as

$$-c_n \Delta_g u + R_g u = R_{\tilde{g}} u^{\frac{n+2}{n-2}}, \quad c_n = \frac{4(n-1)}{n-2}.$$

Therefore, (KW-2) becomes equivalent to finding a positive solution u to

$$-c_n \Delta_g u + R_g u = K u^{\frac{n+2}{n-2}} \quad \text{on } M. \tag{E_K}$$

The linear operator L_g on the left-hand side is the so-called *conformal Laplacian*, and it transforms naturally under conformal changes of metric; see Section 2.2. In two dimensions, the problem instead becomes

$$-\Delta_g w + \frac{1}{2} R_g = \frac{1}{2} R_{\tilde{g}} e^{2w}, \quad \tilde{g} = e^{2w} g.$$

Problem (KW-2) is more rigid than (KW-1), so there are more obstructions to existence, as shown by the following two examples.

Example 1. If R_g has a given sign for $n \geq 3$, then it is impossible to reverse it using conformal deformations; see Chapter 3. This aspect is somehow more similar to the two-dimensional case.

Example 2. There is a well-known obstruction due to Kazdan–Warner on the standard sphere. Suppose u solves (E_K) on $(\mathbb{S}^n, g_{\mathbb{S}^n}) \subseteq \mathbb{R}^{n+1}$, $n \geq 3$, and let $y^1, \dots, y^{n+1}$ be the Euclidean coordinates restricted to $\mathbb{S}^n$. Then one has the identity (see Section A.2)

$$\int_{\mathbb{S}^n} \langle \nabla_{g_{\mathbb{S}^n}} K, \nabla_{g_{\mathbb{S}^n}} y^i \rangle u^{\frac{2n}{n-2}} \, dV_{g_{\mathbb{S}^n}} = 0 \quad \text{for every } i = 1, \dots, n+1.$$

In dimension $n = 2$, the above identity becomes

$$\int_{\mathbb{S}^2} \langle \nabla_{\mathbb{S}^2} K, \nabla_{\mathbb{S}^2} y^i \rangle e^{2w} \, dV_{g_{\mathbb{S}^2}} = 0 \quad \text{for every } i = 1, 2, 3.$$

As a consequence, equation (E_K) is not solvable on $(\mathbb{S}^n, g_{\mathbb{S}^n})$ if K is monotone with respect to any of the Euclidean variables, for example when it is affine. More obstructions to existence are given in [36, 166, 208]; see Chapter 9.

The above non-existence result on $(\mathbb{S}^n, g_{\mathbb{S}^n})$ is due to the presence of the *Möbius group*. This is a non-compact group of conformal maps on $\mathbb{S}^n$, which is obtained by composing the stereographic projection, Euclidean dilations, and then the inverse stereographic map. In this respect, the global maximum of K acts like an *attractor* for the conformal volume; see Chapter 4.

For reasons of brevity we will limit ourselves to the case of curvatures with constant sign: for the changing-sign case we refer the interested reader to the non-exhaustive list [16, 19, 35, 84, 100, 133, 162, 186, 189].

1.3 Variational analysis of (KW-2)

Each conformal class of metrics carries naturally scalar curvatures with a given sign, which is determined by two quantities introduced in Chapter 3. The first is the principal eigenvalue of the conformal Laplacian, $\lambda_1(L_g)$, whose sign is a conformal invariant. The second is the *Yamabe quotient*, defined on conformal factors as

$$Q_g(u) := \frac{\int_M (c_n |\nabla_g u|^2 + R_g u^2) \, dV_g}{\left(\int_M u^{2^*} \, dV_g\right)^{\frac{2}{2^*}}}, \qquad 2^* := \frac{2n}{n-2}.$$

Its infimum, the *Yamabe constant*, is conformally invariant and denoted by $\mathcal{Y}(M, [g])$. It turns out that $\mathcal{Y}(M, [g])$ and $\lambda_1(L_g)$ have the same sign.

In negative curvature it is possible to prove existence of solutions for (KW-2) rather directly using a variational approach. In fact, consider the functional

$$I_K(u) := \frac{1}{2}\int_M (c_n|\nabla_g u|_g^2 + R_g u^2)\,dV_g - \frac{1}{2^*}\int_M K|u|^{2^*}\,dV_g,$$

defined on the space $H^1(M)$ of functions u that are of class $L^2(M)$, together with their gradients. If K is everywhere negative on M, then I_K admits a global minimizer, which is then a solution of (E_K).

For positive curvature the problem is more challenging, as the above Example 2 shows. The energy I_K is in this case unbounded from below, so it is convenient to introduce a modified functional in the form of a Sobolev-type quotient, namely

$$J_K(u) := \frac{\int_M (c_n|\nabla_g u|^2 + R_g u^2)\,dV_g}{\left(\int_M K|u|^{2^*}\,dV_g\right)^{\frac{2}{2^*}}}.$$

Notice that if $\lambda_1(L_g) > 0$, the quadratic form in the numerator is equivalent to the squared norm of $H^1(M)$. The advantage of this formulation is that J_K is bounded from below by the Sobolev embedding of $H^1(M)$ into $L^{2^*}(M)$, and its critical points give rise to solutions of (E_K), after a proper dilation by a positive constant.

However, the latter embedding being non-compact, minimizing sequences may not converge. This phenomenon can be clearly illustrated on the round sphere, especially in relation to the Kazdan–Warner obstruction.

When $K \equiv 1$, J_K on $(\mathbb{S}^n, g_{\mathbb{S}^n})$ reduces to

$$J_{K\equiv 1}(u) := \frac{\int_{\mathbb{S}^n} (c_n|\nabla_g u|^2 + n(n-1)u^2)\,dV_{g_{\mathbb{S}^n}}}{\left(\int_{\mathbb{S}^n} |u|^{2^*}\,dV_{g_{\mathbb{S}^n}}\right)^{\frac{2}{2^*}}}. \tag{1.1}$$

It turns out that on the round sphere the infimum $\mathcal{Y}(\mathbb{S}^n, [g_{\mathbb{S}^n}])$ of the above quotient is equal to $n(n-1)|\mathbb{S}^n|^{\frac{2}{n}}$, and that minimizers can be explicitly classified due to some independent work by Aubin and Talenti; see Chapter 4. Minimizers turn out to be of the form

$$\varphi_{p,\lambda} = \left(\frac{2\lambda}{\lambda^2 + 1 - (\lambda^2 - 1)\cos d_{\mathbb{S}^n}(p,x)}\right)^{\frac{n-2}{2}}, \quad p \in \mathbb{S}^n,\ \lambda > 0. \tag{1.2}$$

Such functions, called *bubbles*, arise as conformal factors of Möbius maps from $\mathbb{S}^n$ to itself: for $\lambda = 1$, $\varphi_{p,\lambda}$ is identically equal to 1, while for λ large the conformal volume density $\varphi_{p,\lambda}^{2^*}$ associated to the corresponding Möbius map concentrates distributionally as a Dirac mass, precisely as $|\mathbb{S}^n|\delta_p$. It also turns out that both the numerator and the denominator in (1.1) are independent of p and λ. It is then possible to prove that J_K on $(\mathbb{S}^n, g_{\mathbb{S}^n})$ has no minimizer unless K is identically constant; see Proposition 4.3.

Extending a result by Moser [178] for the two-dimensional sphere, Escobar and Schoen were able to prove an existence theorem for symmetric functions, stated here in a particular form.

Theorem B ([100]). *Let $n = 3$ and let $K\colon \mathbb{S}^3 \to \mathbb{R}$ be positive and antipodally symmetric. Then* (E_K) *is solvable.*

The strategy relies on working in the class of antipodally symmetric functions, where loss of compactness may only occur with the formation of multiple bubbles at couples of antipodal points. The minimal blow-up energy in a symmetric situation can be computed explicitly as $2^{1-\frac{2}{2^*}}\mathcal{Y}(\mathbb{S}^3,[g_{\mathbb{S}^3}])$; see Chapter 4. Escobar and Schoen proved that for λ large one has

$$J_K(\varphi_{p,\lambda}+\varphi_{-p,\lambda}) < 2^{1-\frac{2}{2^*}}\mathcal{Y}(\mathbb{S}^3,[g_{\mathbb{S}^3}]),$$

showing that minimizing sequences within the symmetric class must stay compact.

1.4 Sub-critical approximation and blow-up analysis

For a given function K not necessarily symmetric, in view of the above-mentioned non-existence of minimizers, one may wonder whether there could be critical points of saddle type for the functional J_K. Such critical points are usually found via min-max schemes or Morse-theoretical tools. When a lack of compactness occurs, as in the present case, these topological tools yield in general *Palais–Smale sequences*, namely sequences of functions along which J_K converges and the gradient of J_K tends to zero. Such sequences were analyzed for a slightly different setting in [207], where it was proved that they *split* into their weak limit and a given number of bubbles as in (1.2) with λ large.

Since there is no information in general about the location or the concentration rates of such bubbles, a useful tool for studying the problem without any symmetry requirement is the sub-critical approximation of (E_K), in the spirit of the well-known paper [192] by Sacks and Uhlenbeck.

For τ positive and small and L_g the conformal Laplacian, we introduce the problem

$$L_g u = K u^{\frac{n+2}{n-2}-\tau}. \qquad (\mathrm{E}_{K,\tau})$$

Its solutions are critical points of the modified Euler–Lagrange functional

$$J_{K,\tau}(u) := \frac{\int_M (c_n|\nabla_g u|^2 + R_g u^2)\, dV_g}{\left(\int_M K u^{2^*-\tau}\, dV_g\right)^{\frac{2}{2^*-\tau}}}.$$

This time, by the compactness of the Sobolev embedding $H^1(M) \to L^{2^*-\tau}(M)$, it is easy to minimize $J_{K,\tau}$, and to find (positive) solutions of $(\mathrm{E}_{K,\tau})$. Furthermore, by a result in [112], for τ positive there is an upper bound (depending on τ) on all positive solutions of this equation.

On the other hand, as $\tau \to 0$, the latter might sometimes blow up. Indeed, this must necessarily happen in situations where the Kazdan–Warner obstruction holds, since there cannot be non-trivial limits in this case. The hope, however, is that some classes of solutions might anyway stay bounded for suitable K, in addition to those that are blowing up.

To understand the diverging behavior one can use the so-called *blow-up analysis* method, which consists in rescaling solutions using the dilation covariance (of either (E_K) or $(\mathrm{E}_{K,\tau})$) in order to obtain bounded positive functions satisfying

$$-c_n \Delta u = K(\bar{x}) u^{\frac{n+2}{n-2}} \quad \text{in } \mathbb{R}^n,$$

where $\bar{x}$ is a blow-up point. Such limiting profiles were classified in [47], and for $K(\bar{x}) = c_n n(n-2)$ are of type

$$U_{\lambda,x_0}(x) := \lambda^{\frac{n-2}{2}} U(\lambda(x - x_0)), \quad \lambda > 0, \ x_0 \in \mathbb{R}^n,$$

where

$$U(x) := \frac{1}{(1 + |x|^2)^{\frac{n-2}{2}}}.$$

If the value of $K(\bar{x})$ is different, it is sufficient to multiply the above expression by a suitable factor, due to the double homogeneity of the equation. Since in the *blow-down* procedure for solutions of $(\mathrm{E}_{K,\tau})$ one can obtain entire solutions in $\mathbb{R}^n$ which are upper bounded by, say, the constant 1, it means that original solutions with large L^∞-norm have the shape of $U_{\lambda,0}(x)$ in suitable coordinates x for some large λ.

The above analysis helps in proving that blow-ups for equation $(\mathrm{E}_{K,\tau})$ are sometimes *isolated simple*, meaning that near each blow-up point there is formation of at most one bubbling profile. This is a rather delicate property that depends on the dimension, on the underlying manifold, and on the structure of the function K, especially on its Taylor-type expansion near its critical points.

There are several results in this direction, some of which are recalled in Chapter 6. The properties that are most relevant for our purposes are the following ones. We consider a sequence $\tau_j \searrow 0$ and a sequence $(u_j)_j$ of positive solutions of (E_{K,τ_j}) (i.e. of $(\mathrm{E}_{K,\tau})$ with $\tau = \tau_j$) with K satisfying

$$K > 0 \text{ is a Morse function such that } \Delta_g K(x) \neq 0 \text{ whenever } \nabla_g K(x) = 0. \quad \text{(ND)}$$

Then the following properties hold:

(i) If $n = 3$ and K satisfies (ND), there is at most one blow-up point, which is isolated simple. If (M^3, g) is not conformally equivalent to $(\mathbb{S}^3, g_{\mathbb{S}^3})$, then there cannot be blow-ups of solutions.

(ii) If $n = 4$ and K satisfies (ND), blow-ups are finitely many and isolated simple.

(iii) If $n \geq 5$ and K satisfies (ND), blow-ups are finitely many and isolated simple provided that the sequence $(u_j)_j$ is uniformly bounded in the norm $H^1(M)$ with zero weak limit.

In each of these three cases, blow-ups for $(u_j)_j$ may only occur at critical points of K with negative Laplacian. More comments for higher dimensions are given in Chapter 10.

The above results are complemented by Theorem 7.1, proven in Chapter 7, in which solutions with isolated simple blow-ups, possibly multiple in dimension $n \geq 4$, are constructed via a refined implicit function argument. One can also characterize the limit values of J_{K,τ_j} on such solutions for j tending to infinity, as well as their Morse index.

In dimension $n = 4$ there are indeed restrictions on the location of blow-up points (see Remark 7.2), while in dimension $n \geq 5$ blow-up may occur at any chosen set among critical points of K with negative Laplacian. If in dimension $n \geq 5$ one removes the above extra assumptions on the Sobolev norm and the weak limit of solutions, then the blow-up behavior might be more involved; see e.g. [62,63]. Exclusion of blow-ups on three-dimensional manifolds differing from the sphere is proved via the *positive mass theorem* by Schoen–Yau; see [198].

1.5 Some general existence results

The above characterizations of blowing-up solutions allow one to derive some general existence results, not requiring any symmetry on K. For example, on $\mathbb{S}^3$ the following theorem holds true, in the spirit of one by Chang and Yang [59] in two dimensions (see also [124]).

Theorem C ([24,57]). *Let $K: \mathbb{S}^n \to \mathbb{R}$ be a positive Morse function satisfying* (ND). *Let $(x_i)_i$ be the (finitely many) critical points of K, with Morse index k_i. Suppose that*

$$\sum_{x_i \text{ s.t. } \Delta K(x_i)<0} (-1)^{k_i} \neq (-1)^n. \qquad (*)$$

If $n = 3$, then (E_K) *is solvable.*

The theorem is proved via a degree-counting formula. By the above results on blow-up analysis, a dichotomy for solutions of $(E_{K,\tau})$ can be shown, namely that they either stay uniformly bounded as $\tau \to 0$ or they develop simple blow-ups at a single point. For τ small, however, it can be shown via a homotopy argument (deforming the function K to 1 in an affine way) and a rigidity result from [34, 112] that the total Leray–Schauder degree of solutions is equal to 1. Subtracting by excision the contribution to the degree from blowing-up solutions, one obtains from the above condition $(*)$ that the uniformly bounded solutions of $(E_{K,\tau})$ contribute non-trivially to the degree formula, proving that they cannot be an empty set.

Remark 1.4. For $n = 4$, a more complicated formula than $(*)$ appears due to the presence of multiple blow-ups: more details are given in Remark 6.16. An existence result on $\mathbb{S}^n$ under condition $(*)$ in any dimension was proved in [5, 61] provided K satisfies a *pinching condition*, that is,

$$\frac{\sup_{\mathbb{S}^n} K}{\inf_{\mathbb{S}^n} K} < p_n,$$

for some constant $p_n > 1$ close enough to 1. In this case we can use a finite-dimensional reduction of the problem related to one presented in Chapter 7, but using degree-theoretical arguments for the reduced problem.

Another recent result requiring pinching conditions is the following one.

Theorem D ([166]). *Suppose $n \geq 5$ and that (M^n, g) is a closed Einstein manifold of positive scalar curvature. Assume $K: M \to \mathbb{R}$ is Morse, satisfies* (ND)*, and that one of the following two conditions is fulfilled:*

(i) $\frac{\sup_M K}{\inf_M K} < 2^{\frac{1}{n-2}}$ *and* $(*)$ *holds;*

(ii) $\frac{\sup_M K}{\inf_M K} < (\frac{3}{2})^{\frac{1}{n-2}}$ *and K has at least two critical points with negative Laplacian.*

Then (E_K) *is solvable.*

Remark 1.5. Some comments are in order:

(i) The second pinching condition in Theorem D is clearly stronger than the first one, but in order for $(*)$ to hold one needs at least two critical points with negative Laplacian. Notice that, even under the strongest pinching conditions, the inequality in $(*)$ is sufficient but not necessary for existence; see also [174]. Under hypothesis (i) the theorem is proven in [61] for $K = 1 + \varepsilon \widetilde{K}$ and in [70] for $(\mathbb{S}^n, g_{\mathbb{S}^n})$.

(ii) The result is false in dimensions $n = 3, 4$.

(iii) As we will see in Chapter 9, assuming that only one critical point of K has negative Laplacian is not sufficient in general to guarantee existence of solutions, even if one imposes stronger pinching conditions.

The latter result extends a theorem from [70] which is valid for the sphere under assumption (i). Recall that Einstein means that $\mathrm{Ric}_g = \Lambda g$ for some $\Lambda \in \mathbb{R}$. Also in this case, the rigidity results from [112] and [34] apply, giving that the total degree of the solutions of $(\mathrm{E}_{K,\tau})$ is equal to 1.

Differently from the three-dimensional case though, blow-ups may now occur at multiple critical points of K. However, the blow-up analysis presented before still allows one to compute their contribution to the degree *within some given levels* of $J_{K,\tau}$ that include single- and double-bubbling solutions. We can then conclude the proof by applying the classical *Poincaré–Hopf theorem*.

In Chapter 9 we give examples of arbitrarily pinched curvature functions with only one critical point of negative Laplacian and arbitrary Morse structure such that (E_K) is not solvable, as well as non-existence examples in dimensions $n = 3$ or 4 for functions with two critical points of negative Laplacian. These show the sharpness of the assumptions in Theorem D.

We conclude the notes with some perspectives and a set of open problems. These concern extensions of the above-mentioned blow-up analysis, the possible use of Morse homology, higher-order and fully non-linear versions of the Kazdan–Warner problem, complete manifolds, manifolds with boundary, and the role of conformal geometry in the study of Einstein's constraint equations.

Chapter 2
Preliminaries

In this chapter we recall some useful preliminary facts about Riemannian manifolds, properties of certain functions spaces in the closed case, and some regularity theory for elliptic equations, both of linear and of non-linear type.

2.1 A quick review of differential geometry

Let M be an n-dimensional smooth closed manifold (compact and without boundary). Under such conditions, M is covered by a finite family of charts (an *atlas*) $(U_i, \varphi_i)_{i=1,\dots,N}$ with $U_i \subseteq M$ open for $i = 1, 2, \dots, N$, such that

$$\varphi_i : U_i \to V_i \text{ is a homeomorphism,}$$

where $V_i \subseteq \mathbb{R}^n$ is an open set too. One also requires that for every $i, j = 1, 2, \dots, N$, if $U_i \cap U_j \neq \emptyset$, the *transition map* (see Figure 2.1)

$$\varphi_i \circ \varphi_j^{-1} : \varphi_j(U_i \cap U_j) \subseteq \mathbb{R}^n \to \varphi_i(U_i \cap U_j) \subseteq \mathbb{R}^n$$

is smooth; see e.g. [95]. If (U, φ) is a chart around a point $p \in M$, then $x^i := y^i \circ \varphi$, $i = 1, 2, \dots, n$, are called *local coordinates* on U, where the y^i are standard Euclidean coordinate functions on $\mathbb{R}^n$.

Definition 2.1. Let M_1 and M_2 be two smooth manifolds. A function $f : M_1 \to M_2$ is said to be *differentiable* (respectively, *of class* C^k) at a point $p \in M$ if (U_1, φ_1) and (U_2, φ_2) are charts around p and $f(p)$ respectively, then the map $\varphi_2 \circ f \circ \varphi_1^{-1}$ is differentiable at $\varphi_1(p)$ (respectively, of class C^k).

With this definition in hand, we can define tangent vectors at a point $p \in M$.

Definition 2.2. Let $\gamma : (-\varepsilon, \varepsilon) \to M$ be a differentiable curve with $\gamma(0) = p$. Let $C^\infty(M)$ be the set of smooth functions on M, namely that are smooth when expressed in coordinates. The linear map

$$\gamma'(0) : C^\infty(M) \to \mathbb{R}, \quad \gamma'(0) f := \frac{d}{dt}(f \circ \gamma)|_{t=0}$$

is said to be the *tangent vector to the curve* γ at $t = 0$. A *tangent vector* to M at p is the tangent vector at $t = 0$ of some curve $\gamma : (-\varepsilon, \varepsilon) \to M$ as above with $\gamma(0) = p$.

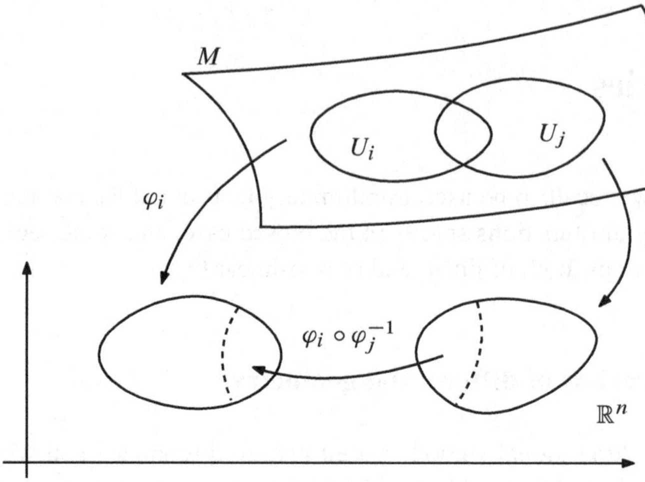

Figure 2.1. Transition map between charts

The set of all tangent vectors at p is denoted by T_pM and has the structure of a vector space. The set

$$TM := \{(p, v) \colon p \in M,\ v \in T_pM\}$$

is called the *tangent bundle of* M, and it naturally inherits a differentiable structure from that of M.

If the dimension of M is n, then T_pM is an n-dimensional vector space. Moreover, in local coordinates x^i, the set $\{\frac{\partial}{\partial x^1}|_p, \dots, \frac{\partial}{\partial x^n}|_p\}$ forms a basis of T_pM, where

$$\frac{\partial}{\partial x^i}\Big|_p f := \frac{\partial}{\partial y^i}(\varphi \circ f)|_{\varphi(p)},$$

with y^i as above.

If $f \colon M_1 \to M_2$ is a smooth map, then the *differential* of f at $p \in M$ is defined as a linear map from T_pM_1 to $T_{f(p)}M_2$ as follows. For a given tangent vector $V \in T_pM_1$, let $\gamma \colon (-\varepsilon, \varepsilon) \to M_1$ be a curve with $\gamma(0) = p$ and $\gamma'(0) = v$. Then $f \circ \gamma$ is a curve in M_2 with $f \circ \gamma(0) = f(p)$, and we define $df_p(v) := (f \circ \gamma)'(0) \in T_{f(p)}M_2$.

Definition 2.3. A *vector field* X on a manifold is a map from M into TM such that $X(p) \in T_pM$ for all $p \in M$. If the map $p \mapsto X(f)(p) := X(p)f$ is smooth for every smooth function f on M, then X is said to be a *smooth vector field.*

Let $\chi(M)$ denote the family of smooth vector fields on M.

Definition 2.4. An *affine connection* D on a manifold M is a map

$$D\colon \chi(M) \times \chi(M) \to \chi(M), \quad (X, Y) \mapsto D_X Y,$$

which satisfies the following properties for every $X, Y, Z \in \chi(M)$ and all $f, g \in \mathcal{D}(M)$:

(i) $D_X(Y + Z) = D_X Y + D_X Z$;

(ii) $D_X(fY) = f D_X Y + X(f)Y$;

(iii) $D_{fX+gY} Z = f D_X Z + g D_X Z$.

A *Riemannian manifold* (M, g) is a manifold M equipped with a symmetric positive-definite smooth tensor g of type $(2, 0)$ (twice covariant), namely a bilinear form on $T_p M$ for every $p \in M$. In local coordinates $(x^i)_i$, with the standard convention of summing over repeated upper and lower indices, the metric g can be expressed as $g = g_{ij}\, dx^i \otimes dx^j$, where $g_{ij} = g_{ji}$, and for any two tangent vectors $V, W \in T_p M$ we have

$$\begin{aligned} g(V, W) &= \langle V, W \rangle_g \\ &= \sum_{i,j=1}^{n} g_{ij} V^i W^j, \end{aligned}$$

where $(V^i)_i$ and $(W^j)_j$ are the components of V and W with respect to the coordinates $(x^i)_i$, namely expressed as

$$V = V^i \frac{\partial}{\partial x^i}\Big|_p, \quad W = W^j \frac{\partial}{\partial x^j}\Big|_p.$$

We have the following well-known result.

Theorem 2.5. *On a smooth Riemannian manifold (M, g) there exists a unique affine connection D (known as Riemannian connection or Levi-Civita connection) satisfying the following conditions:*

(i) *D is symmetric, namely*

$$D_X Y - D_Y X = [X, Y] := XY - YX \quad \textit{for every } X, Y \in \chi(M).$$

(ii) *D is compatible with the metric g, i.e.*

$$X\langle Y, Z \rangle_g = \langle D_X Y, Z \rangle_g + \langle Y, D_X Z \rangle_g,$$

for every $X, Y, Z \in \chi(M)$.

In local coordinates $x = (x^1, \dots, x^n)$, for the *coordinate vector fields* $X_i := \frac{\partial}{\partial x^i}$ and $X_j := \frac{\partial}{\partial x^j}$ we can write

$$D_{X_i} X_j = \sum_{k=1}^{n} \Gamma_{ij}^k \frac{\partial}{\partial x^k}.$$

The coefficient functions Γ_{ij}^k are called the *Christoffel symbols* of the Riemannian connection D. It can be proved that they are determined by the metric g through the formula

$$\Gamma_{ij}^k = \frac{1}{2} \sum_{\ell=1}^{n} \Big(\frac{\partial}{\partial x^i} g_{j\ell} + \frac{\partial}{\partial x^j} g_{i\ell} - \frac{\partial}{\partial x^\ell} g_{ij} \Big) g^{k\ell},$$

where $(g^{ij})_{ij}$ is the inverse of the matrix of coefficients $g = (g_{ij})_{ij}$. The symmetry property of D implies that $\Gamma_{ij}^k = \Gamma_{ji}^k$ for every $i, j, k = 1, \dots, n$.

Definition 2.6. The *Riemann curvature* $R\colon \chi(M)^3 \to \chi(M)$ is defined by

$$R(X, Y)Z = D_X D_Y Z - D_Y D_X Z - D_{[X,Y]} Z$$

for every vector field X, Y, Z, where D is the Riemannian connection and $[\cdot, \cdot]$ the Lie bracket. It can be shown that $(R(X, Y)Z)(p)$ depends only on the values of X, Y, and Z at p, i.e. R is a *tensorial object.*

The curvature tensor R satisfies the following identities for all $X, Y, Z, W \in \chi(M)$:

(i) $R(X, Y)Z = -R(Y, X)Z$;

(ii) $R(X, Y)Z + R(Y, Z)X + R(Z, X)Y = 0$ (Bianchi's identity);

(iii) $\langle R(X, Y)Z, W \rangle_g = -\langle R(X, Y)W, Z \rangle_g$;

(iv) $\langle R(X, Y)Z, W \rangle_g = -\langle R(Z, W)X, Y \rangle_g$.

Definition 2.7. The *Ricci curvature* Ric is the symmetric 2-tensor

$$\mathrm{Ric}(X, Y) := \text{trace of the linear map } Z \mapsto R(X, Z)Y.$$

The *scalar curvature* is defined as the trace of the Ricci curvature tensor with respect to g, i.e. $g^{ij} \mathrm{Ric}_{ij}$. As before, we are adopting the standard convention of summing from 1 to n when using repeated upper and lower indices. Since we will rarely employ the Riemann tensor, we will use R_g to denote the scalar curvature, being confident that this will not generate confusion.

It will also be useful to consider the *Weyl tensor*, a tensor of type $(4, 0)$ defined as

$$W_g = R - \frac{1}{n-2} \Big(\mathrm{Ric} - \frac{R_g}{n} g \Big) \owedge g - \frac{R_g}{2n(n-1)} g \owedge g.$$

In this formula $\owedge$ stands for the Kulkarni–Nomizu product of symmetric $(2,0)$ tensors, which is defined by

$$(h \owedge k)(X,Y,Z,W) := h(X,Z)k(Y,W) + h(Y,W)k(X,Z) \\ - h(X,W)k(Y,Z) - h(X,Z)k(X,W). \quad (2.1)$$

The Weyl tensor represents the *conformal part* of the Riemann tensor, and we will show below how it transforms under a conformal change of metric.

If (M,g) is a two-dimensional manifold, for example a surface immersed in $\mathbb{R}^3$, the Gaussian curvature of g is exactly one half of the scalar curvature R_g.

We recall that a Riemannian manifold (M,g) is said to be *Einstein* if there exists a constant Λ such that

$$\mathrm{Ric}_g(X,Y) = \Lambda g(X,Y) \quad \text{for every } X,Y \in \chi(M).$$

Definition 2.8. Let f be a smooth function on a Riemannian manifold (M,g). The *gradient* $\nabla_g f(p)$ of f with respect to the metric g at $p \in M$ is an element of T_pM, and is defined by the relation

$$\langle \nabla_g f(p), v\rangle_g = df_p(v) \quad \text{for every } v \in T_pM.$$

In local coordinates,

$$\nabla_g f = g^{ij} \frac{\partial f}{\partial x^j} \frac{\partial}{\partial x^i}.$$

Definition 2.9. The *Hessian* $\nabla^2 f$ of a smooth function f is a $(2,0)$ tensor field defined as

$$\nabla^2 f(X,Y) = XYf - D_X Yf.$$

It is symmetric, and its trace is called the *Laplacian*, denoted by $\Delta_g f$.

In local coordinates, setting $\partial_i := \frac{\partial}{\partial x^i}$, one has the expression

$$\Delta_g f = \frac{1}{\sqrt{|g|}} \sum_{i,j=1}^{n} \partial_i (g^{ij} \sqrt{|g|} \partial_j f), \quad |g| := \det(g_{ij}).$$

The Laplace operator Δ_g is self-adjoint on a compact manifold (M,g) without boundary. Moreover, it satisfies the integration by parts formula

$$\int_M f_1 \Delta_g f_2 \, dV_g = \int_M f_2 \Delta_g f_1 \, dV_g = -\int_M g(\nabla f_1, \nabla f_2)\, dV_g,$$

for every $f_1, f_2 \in C^\infty(M)$. To verify this, one can use the fact that

$$dV_g = \sqrt{|g|}\, dx, \quad g(\nabla f_1, \nabla f_2) = g^{ij} \partial_i f_1 \partial_j f_2,$$

and the representation of $\Delta_g f$ in local coordinates. Conversely, if the above integration by parts formula were to hold, then the local coordinate expression for Δ_g would necessarily have to be as above.

2.2 Conformal changes of metric

Definition 2.10. Let (M, g) be a smooth Riemannian manifold. A metric $\tilde{g}$ on M is said to be *conformal* to g if there exists a function $u \in C^\infty(M)$ such that $\tilde{g} = e^{2u} g$. The set of all metrics $\tilde{g}$ conformal to g is said to be the *conformal class* of g, and is denoted by $[g]$.

We recall below several relations between geometric objects with respect to the metrics g and $\tilde{g} = e^{2u} g$ (see e.g. [32]):

(i) Ricci tensor:

$$\mathrm{Ric}_{\tilde{g}} = \mathrm{Ric}_g - (n-2)(\nabla_g^2 u - du \otimes du) - (\Delta_g u + (n-2)|du|_g^2) g.$$

(ii) Scalar curvature:

$$R_{\tilde{g}} = e^{-2u}\big(R_g - 2(n-1)\Delta_g u - (n-2)(n-1)|du|_g^2\big).$$

(iii) Volume form:

$$dV_{\tilde{g}} = e^{nu}\, dV_g.$$

(iv) Laplacian:

$$\Delta_{\tilde{g}} f = e^{-2u}(\Delta_g f - (n-2) g(du, df)).$$

(v) Weyl tensor:

$$W_{\tilde{g}} = e^{2u} W_g.$$

Notice that the operator $-\Delta_g$ is conformally covariant in dimension $n = 2$, i.e. with $\tilde{g} = e^{2u} g$ we have

$$(-\Delta_{\tilde{g}}) f = e^{-2u}(-\Delta_g) f \quad \text{for every } f \in C^\infty(M).$$

In dimension $n \geq 3$ one needs to add a multiple of the scalar curvature R_g to obtain a similar property: setting

$$L_g := -c_n \Delta_g + R_g, \quad c_n := \frac{4(n-1)}{n-2}, \tag{2.2}$$

then for $\tilde{g} = u^{\frac{4}{n-2}} g$ we have

$$L_{\tilde{g}} f = u^{-\frac{n+2}{n-2}} L_g(uf) \quad \text{for every } f \in C^\infty(M). \tag{2.3}$$

The operator L_g is known as the *conformal Laplacian* and it is self-adjoint. Notice that by the above formulas and a change of variables one finds

$$R_{\tilde{g}} = u^{-\frac{n+2}{n-2}} L_g(u), \tag{2.4}$$

for any positive and smooth $u\colon M \to \mathbb{R}$.

2.3 Hölder and Sobolev spaces on closed manifolds

In order to understand the mapping properties of the Laplace–Beltrami operator Δ_g, or of the conformal Laplacian L_g, we need to introduce Hölder and Sobolev spaces on closed manifolds (see e.g. the books [129–131]).

Given $\alpha \in (0, 1)$, the space $C^\alpha(M)$ is the family of continuous functions on M such that

$$\|u\|_{C^\alpha(M)} := \max_M |u| + \sup_{x \neq y} \frac{|u(x) - u(y)|}{d_g(x, y)^\alpha} < \infty.$$

We recall that for $u\colon M \to \mathbb{R}$ smooth, ∇u is its gradient, a vector field with components $(\nabla u)^i = g^{ij} \partial_j u$. The right notion of a higher-order derivative on a manifold is the *kth covariant derivative* $\nabla^k u$, i.e. a multi-derivative of order k with lower-order correction terms (well behaved under changes of coordinates). For example, if $k = 2$ it coincides with the Hessian and in local coordinates its components are

$$(\nabla^2 u)_{ij} = \partial^2_{ij} u - \Gamma^k_{ij} \partial_k u,$$

where the Γ^k_{ij} are the Christoffel symbols. With these corrections, $\nabla^k u$ becomes a *tensor field*. The squared norm of $\nabla^k u$ is

$$|\nabla^k u|^2 := g^{i_1 j_1} \cdots g^{i_k j_k} (\nabla^k u)_{i_1,\ldots,i_k} (\nabla^k u)_{j_1,\ldots,j_k}.$$

For $k \geq 1$ and $u \in C^\infty(M)$ set

$$\|u\|_{W^{k,p}(M)} := \sum_{j=0}^{k} \Big(\int_M |\nabla^j u|^p \, dV_g \Big)^{\frac{1}{p}}, \quad p \geq 1,$$

and define $W^{k,p}(M)$ to be the completion of $C^\infty(M)$ with respect to this norm.

On (M, g) closed, one can also define higher-order Hölder norms in the following way: cover M with finitely many charts $(U_i, \varphi_i)_{i=1,\ldots,N}$, so M is also covered by the interior of smooth sets (V_i) compactly supported in U_i. For k a non-negative integer and $\alpha \in (0, 1)$ we then set

$$\|u\|_{C^{k,\alpha}}(M) := \sum_{i=1}^{N} \|u \circ \varphi_i^{-1}\|_{C^{k,\alpha}(\bar{V}_i)}.$$

Remark 2.11. The space $H^k(M) := W^{k,2}(M)$ is Hilbert, and $W^{k,p}(M)$ is reflexive if $1 < p < \infty$.

Theorem 2.12. *Let (M^n, g) be closed. Then for all $1 \leq q < p$ and integers $0 \leq m \leq k$ such that*

$$\frac{1}{p} = \frac{1}{q} - \frac{k-m}{n},$$

one has the Sobolev embedding $W^{k,q} \hookrightarrow W^{m,p}(M)$. If $q \geq 1$ and $\alpha \in (0,1)$ with

$$\frac{1}{q} \leq \frac{1-\alpha}{n},$$

then $W^{1,q}(M) \hookrightarrow C^\alpha(M)$ (Morrey's embedding).

2.4 Regularity and existence theory for linear elliptic operators

We next state some standard regularity results for linear operators of Schrödinger type, namely sums of the Laplacian and a potential term.

Theorem 2.13. *Let (M^n, g) be closed, and let $a\colon M \to \mathbb{R}$ be smooth. Then we have the following properties:*

(i) *For every $\alpha \in (0,1)$ there exists $C = C(M, g, a, \alpha)$ such that if $u \in C^2(M)$ solves*

$$(-\Delta_g + a)u = f,$$

with $f \in C^\alpha(M)$, then

$$\|u\|_{C^{2,\alpha}(M)} \leq C(\|f\|_{C^\alpha(M)} + \|u\|_{L^\infty(M)}).$$

Moreover, if $f \in C^{k,\alpha}(M)$ for some $k \geq 1$, then there exists $\widetilde{C} = \widetilde{C}(M, g, a, k, \alpha)$ such that

$$\|u\|_{C^{k+2,\alpha}(M)} \leq \widetilde{C}(\|f\|_{C^{k,\alpha}(M)} + \|u\|_{L^\infty(M)}).$$

(ii) *If $p > 1$ and $k \geq 1$ an integer, then there exists $C = C(M, g, p, k, a)$ such that if $f \in W^{k,p}(M)$ and $u \in W^{k+2,p}(M)$ solves*

$$(-\Delta_g + a)u = f \quad \text{a.e. on } M,$$

then

$$\|u\|_{W^{k+2,p}(M)} \leq C(\|f\|_{W^{k,p}(M)} + \|u\|_{L^p(M)}).$$

Notice that $(-\Delta_g + a)$ is self-adjoint. Using Fredholm's theory one can prove the following theorem:

Theorem 2.14. *Suppose that* $(-\Delta_g + a)$ *has trivial kernel. Then, in the setting of the latter theorem, solutions exist and the norms* $\|u\|_{\cdot}$ *can be removed from the right-hand sides.*

Remark 2.15. Weaker norms of u on the right-hand side are needed in the case that the operator $(-\Delta_g + a)$ is not invertible, as one can see by choosing a non-trivial element of the kernel as u.

Remark 2.16. We notice the following:

(i) In the case a of trivial kernel, solutions are represented via a Green's function (see e.g. [16]) $G\colon M \times M \setminus \text{diagonal} \to \mathbb{R}$, with the same singularity behavior as in $\mathbb{R}^n$ and smooth away from the diagonal.

(ii) The operator $(-\Delta_g + a)$ has a sequence of eigenvalues $\lambda_1 < \lambda_2 \leq \lambda_3 \cdots \to +\infty$. The first eigenvalue λ_1 is simple with positive eigenfunction φ_1, as one can see using its characterization via the Rayleigh quotient. If $\lambda_1 > 0$, then $(-\Delta_g + a)$ is invertible, it satisfies the maximum principle (i.e. $f \geq 0$, $f \not\equiv 0 \Rightarrow u > 0$) and the Green's function is positive.

2.5 Regularity of weak solutions to some non-linear equations

Combining the above linear regularity theory and Sobolev's embeddings, we next prove smoothness of weak solutions to equations of the form

$$-\Delta_g u + a(x)u = K(x)|u(x)|^{p-1}u(x) \quad \text{on } M, \quad p \in \left(1, \tfrac{n+2}{n-2}\right]. \tag{E_p}$$

Here, $a, K\colon M \to \mathbb{R}$ are smooth functions, and the equation includes (E_K) and $(\mathrm{E}_{K,\tau})$ as particular cases, allowing changing-sign solutions also. We say that $u \in H^1(M)$ is a *weak solution* of (E_p) if it satisfies

$$\int_M (\langle \nabla_g u, \nabla_g v\rangle + a(x)uv - K(x)|u|^{p-1}uv)\, dV_g = 0 \quad \text{for all } v \in H^1(M).$$

Notice that the last term in the integral is of class $L^1(M)$ by Theorem 2.12 and Hölder's inequality. Below we will prove regularity of weak solutions to (E_p), first recalling a couple of useful facts.

Fact 1 ([113, Theorem 8.8]). Suppose $u \in H^1(M)$ is a weak solution to

$$-\Delta_g u + a(x)u = f,$$

where a is smooth and $f \in L^2(M)$. This means that

$$\int_M (\langle \nabla_g u, \nabla_g v\rangle + a(x)uv - fv)\, dV_g = 0 \quad \text{for every } v \in H^1(M).$$

Then $u \in W^{2,2}(M)$ and it is a *strong solution*, namely it satisfies the equation a.e. in M.

Since $u \in H^1(M) \hookrightarrow L^{\frac{2n}{n-2}}(M)$ and $p \leq \frac{n+2}{n-2}$, the right-hand side belongs to $L^{\frac{2n}{n+2}}(M)$, therefore not to $L^2(M)$ in general. We can anyway recover the previous property for less integrable right-hand sides using an approximation procedure.

Fact 2. Suppose $u \in H^1(M)$ weakly solves

$$-\Delta_g u + a(x)u = f \in L^{\frac{2n}{n+2}}(M).$$

Then $u \in W^{2,\frac{2n}{n+2}}(M)$ and it is a strong solution.

To prove this fact we consider a sequence of smooth functions $(f_k)_k$ such that

$$f_k \to f \text{ in } L^{\frac{2n}{n+2}}(M), \qquad \int_M f_k \, dV_g = \int_M f \, dV_g,$$

and let u_k solve

$$-\Delta_g u_k = f_k - a(x)u.$$

Notice that

$$0 = \int_M \Delta_g u \, dV_g = \int_M (au - f) \, dV_g \quad \Rightarrow \quad \int_M (au - f_k) \, dV_g = 0,$$

and the solution u_k exists by Fredholm's theory. Then $u_k \in W^{2,2}(M)$ by the previous fact, and by Theorem 2.13 we have $u_k \to u$ in $W^{2,\frac{2n}{n+2}}(M)$.

Hence weak solutions of (E_p) belong to $W^{2,\frac{2n}{n+2}}(M)$.

Theorem 2.17. *Let (M, g) be compact, and let $u \in H^1(M)$ be a weak solution to equation* (E_p). *Then $u \in C^{2,\alpha}(M)$ for all $\alpha \in (0, 1)$.*

Proof. If $p < \frac{n+2}{n-2}$, one can easily obtain regularity of u using *bootstrap*, i.e. combining $W^{2,q}$-estimates with Sobolev's embedding.

In fact, since $u \in L^{r_1}(M)$, $r_1 = 2^*$, the right-hand side is of class $L^{\frac{r_1}{p}}$, and therefore by Fact 2, u is of class $W^{2,\frac{r_1}{p}}$. Assuming that this space does not embed into Hölder continuous functions, by Sobolev's embedding u is anyway of class

$$L^{\frac{n\frac{r_1}{p}}{n-2\frac{r_1}{p}}}.$$

We then need to check whether

$$\frac{n\frac{r_1}{p}}{n-2\frac{r_1}{p}} > r_1, \quad \text{i.e.} \ \frac{nr_1}{np-2r_1} > r_1,$$

which is true since $np - 2r_1 < n$. Then setting, iteratively,

$$r_{k+1} = \frac{\frac{r_k}{p}}{n - 2\frac{r_k}{p}},$$

one can verify that $W^{2,\frac{r_k}{p}}(M)$ embeds into $C^\alpha(M)$ for some $\alpha \in (0,1)$ and some k large enough. The statement then follows from Theorem 2.13 and Ascoli's theorem.

For $p = \frac{n+2}{n-2}$ there is no improvement of integrability from the above strategy. One way out of this problem is to use a Moser iteration scheme, as recalled e.g. in [209, Appendix B]. We present here an alternative argument from [213], which has the advantage of adapting to higher-order equations as well.

First, we assume that the operator $-\Delta_g + a(x)$ has trivial kernel, so one can apply Theorem 2.14 above. We write

$$-\Delta_g u + au = K(x)|u|^{p-1}u = b(x)u, \quad b := K(x)|u|^{p-1} \in L^{\frac{n}{2}}(M),$$

and set

$$b = b_j + \tilde{b}_j, \quad b_j := b\chi_{|u|\leq j}, \quad \tilde{b}_j = b\chi_{|u|>j}, \quad j \in \mathbb{N}.$$

Notice that $\|\tilde{b}_j\|_{L^{\frac{n}{2}}(M)} \to 0$ as $j \to \infty$, and that $b_j u \in L^\infty(M)$. For every $q > 1$ the operator $-\Delta_g + a$ is invertible from $W^{2,q}(M)$ to $L^q(M)$, by Theorem 2.14. Then write (E_p) as

$$L_j u := -\Delta_g u + au - \tilde{b}_j u = b_j u.$$

We take $q < \frac{n}{2}$ close to $\frac{n}{2}$, so that $u \in W^{2,q}(M)$ implies $u \in L^r(M)$ for r large. By Hölder's inequality, $\tilde{b}_j u \in L^q(M)$ and $\|\tilde{b}_j u\|_{L^q(M)} = o_j(1)\|u\|_{W^{2,q}(M)}$, hence also L_j is invertible and $u \in W^{2,q}(M)$. Notice that $q > \frac{2n}{n+2}$ if q is close to $\frac{n}{2}$, so we indeed improve the initial $W^{2,\frac{2n}{n+2}}$-estimate, and we can now bootstrap to C^α regularity.

If $-\Delta_g + a(x)$ is not invertible, it is sufficient to add a large positive constant in front of u on both sides of the equation, and to apply the same argument to the modified invertible operator on the left-hand side. ∎

Chapter 3

The Yamabe quotient and resolution of (KW-1)

In this chapter we derive some properties of the first eigenvalue of the conformal Laplacian L_g and introduce the Yamabe constant. We will then characterize its value for the standard sphere, discuss the Yamabe conjecture, and prove Theorem A.

3.1 On the first eigenvalue of L_g and the Yamabe quotient

In this section we show that the sign of the first eigenvalue $\lambda_1(L_g)$ of the conformal Laplacian L_g is invariant under conformal changes of the metric.

Proposition 3.1. *Let (M^n, g), $n \geq 3$ be a closed manifold. Then, within the conformal class of g, there exists a metric $\tilde{g}$ with $R_{\tilde{g}} > 0$ (respectively, $= 0$, < 0) if and only if $\lambda_1(L_g) > 0$ (respectively, $= 0$, < 0).*

Proof. Recall from Remark 2.16 that $\lambda_1(L_g)$ is simple; let φ_1 be the corresponding eigenfunction, which we can choose strictly positive by the maximum principle, and let $g_1 := \varphi_1^{\frac{4}{n-2}} g$. Then

$$R_{g_1} = \varphi_1^{-\frac{n+2}{n-2}} L_g(\varphi_1) = \lambda_1(L_g)\varphi_1^{-\frac{n+2}{n-2}} \varphi_1,$$

so R_{g_1} has the sign of $\lambda_1(L_g)$. Vice versa, suppose that $R_{\tilde{g}}$ has constant sign for some conformal metric $\tilde{g} := u^{\frac{4}{n-2}} g$, $u > 0$. Then we multiply the equation

$$L_g u = R_{\tilde{g}} u^{\frac{n+2}{n-2}}$$

by φ_1 and, integrating by parts, pass L_g to φ_1:

$$\int_M \varphi_1 R_{\tilde{g}} u^{\frac{n+2}{n-2}}\, dV_g = \int_M \varphi_1 L_g u\, dV_g = \int_M u L_g \varphi_1\, dV_g = \lambda_1(L_g) \int_m u \varphi_1\, dV_g.$$

Hence $R_{\tilde{g}}$ has the same sign as $\lambda_1(L_g)$. ∎

Corollary 3.2. *One has the following consequences:*

(i) *The sign of $\lambda_1(L_g)$ is conformally invariant.*

(ii) *Starting from a scalar curvature R_g of constant sign, it is not possible to change its sign with conformal changes of the metric.*

Definition 3.3. Conformal classes of metrics such that $\lambda_1(L_g) > 0$ (respectively, $= 0$ or < 0) are said to be of *positive* (respectively, *null* or *negative*) *Yamabe class*.

We will introduce next a more geometric characterization of this sign. Let (M^n, g) be closed with $n \geq 3$. For $0 \not\equiv u \in H^1(M)$ define

$$Q_g(u) := \frac{\int_M (c_n |\nabla_g u|^2 + R_g u^2)\, dV_g}{\left(\int_M |u|^{2^*}\, dV_g\right)^{\frac{2}{2^*}}}, \qquad 2^* := \frac{2n}{n-2} > 2.$$

Notice that the denominator in the above expression is well defined by the Sobolev embedding $H^1(M) \hookrightarrow L^{2^*}(M)$, and the numerator for smooth functions can be written as

$$\int_M u L_g u \, dV_g.$$

The *Yamabe quotient* $\mathcal{Y}(M, g)$ is then defined by

$$\mathcal{Y}(M, g) := \inf_{0 \not\equiv u \in H^1(M)} Q_g(u).$$

The following result holds true.

Theorem 3.4. *The Yamabe quotient $\mathcal{Y}(M, g)$ is finite; it only depends on the conformal class $[g]$ of g (so from now on we will write $\mathcal{Y}(M, g) = \mathcal{Y}(M, [g])$), and it has the same sign as $\lambda_1(L_g)$.*

Proof. Since $c_n > 0$, R_g is smooth, and $2^* > 2$, by Hölder's inequality we obtain

$$\int_M |R_g| u^2 \, dV_g \leq C \left(\int_M |u|^{2^*} \, dV_g \right)^{\frac{2}{2^*}} \quad \Rightarrow \quad Q_g(u) \geq -\widetilde{C},$$

for some $\widetilde{C} > 0$.

Now consider $\tilde{g} = v^{\frac{4}{n-2}} g$, with v smooth and positive on M, and notice that by density of $C^\infty(M)$ in $H^1(M)$ we have

$$\mathcal{Y}(M, g) = \inf_{0 \not\equiv u \in C^\infty(M)} Q_g(u).$$

It also holds that $dV_{\tilde{g}} = v^{2^*} dV_g$, which implies

$$\int_M |u|^{2^*} \, dV_g = \int_M \left| \frac{u}{v} \right|^{2^*} dV_{\tilde{g}}.$$

By the conformal covariance of L_g (see (2.3)) we have

$$\frac{u}{v} L_{\tilde{g}} \left(\frac{u}{v} \right) dV_{\tilde{g}} = \frac{u}{v} v^{-\frac{n+2}{n-2}} L_g(u) v^{2^*} \, dV_g = u L_g u \, dV_g.$$

Hence $Q_g(u) = Q_{\tilde{g}}(\frac{u}{v})$, which implies $\mathcal{Y}(M, g) = \mathcal{Y}(M, \tilde{g}) = \mathcal{Y}(M, [g])$.

Let us now check the sign property: we have by the characterization of $\lambda_1(L_g)$ via Rayleigh's quotient,

$$\lambda_1(L_g) = \inf_{0 \not\equiv u \in H^1(M)} \frac{\int_M (c_n |\nabla_g u|^2 + R_g u^2)\, dV_g}{\int_M u^2\, dV_g}.$$

This implies

$$\lambda_1(L_g) < 0 \text{ if and only if } \mathcal{Y}(M,[g]) < 0.$$

If $\lambda_1(L_g) = 0$, the numerator is non-negative, and hence $\mathcal{Y}(M,[g]) \geq 0$. However, $Q_g(\varphi_1) = 0$ (with φ_1 the first eigenfunction of L_g), therefore

$$\lambda_1(L_g) = 0 \Rightarrow \mathcal{Y}(M,[g]) = 0.$$

On the other hand, if $\mathcal{Y}(M,[g]) = 0$ then we have $\lambda_1(L_g) \geq 0$. But if $\lambda_1(L_g) > 0$, the quadratic form in the numerator is positive definite and equivalent to $\|u\|^2_{H^1(M)}$, so we would have $\mathcal{Y}(M,[g]) > 0$ by Sobolev's inequality. Thus

$$\mathcal{Y}(M,[g]) = 0 \text{ if and only if } \lambda_1(L_g) = 0.$$

The above two equivalences imply that

$$\mathcal{Y}(M,[g]) > 0 \text{ if and only if } \lambda_1(L_g) > 0.$$

This concludes the proof. ■

Remark 3.5. Some comments are in order:

(i) Notice that $Q_g(u) = Q_g(|u|)$, so by density of smooth functions in $H^1(M)$,

$$\mathcal{Y}(M,[g]) = \inf_{0 < u \in C^\infty(M)} Q_g(u).$$

For $u > 0$ smooth, we can write the numerator in Q_g as

$$\int_M u u^{\frac{n+2}{n-2}} R_{\tilde g}\, dV_g = \int_M R_{\tilde g}\, dV_{\tilde g}, \quad \tilde g := u^{\frac{4}{n-2}} g.$$

Therefore,

$$\mathcal{Y}(M,[g]) = \inf_{\tilde g \in [g]} \frac{\int_M R_{\tilde g}\, dV_{\tilde g}}{(\mathrm{Vol}_{\tilde g}(M))^{\frac{2}{2^*}}}.$$

(ii) Critical points of Q_g solve the Euler–Lagrange equation

$$L_g u = \mu u^{\frac{n+2}{n-2}}$$

with a Lagrange multiplier μ. Therefore, if $u > 0$, these are *Yamabe metrics*, namely with constant scalar curvature. The map

$$g \mapsto \int_M R_g \, dV_g$$

is the so-called *Einstein–Hilbert functional*: its stationary points (not restricted to $[g]$) are Einstein metrics. The above action is a strongly indefinite functional: its critical points have infinitely many stable/unstable directions (on the round sphere, stable directions correspond to conformal deformations and unstable ones correspond to non-conformal deformations; see [195]).

The case of $(\mathbb{S}^n, g_{\mathbb{S}^n})$ is particularly important, and we have a comparison to $\mathbb{R}^n$, for which we recall the Sobolev–Gagliardo–Nirenberg inequality (see e.g. [45]):

$$\|u\|_{L^{p^*}(\mathbb{R}^n)} \leq C_{p,n} \|\nabla u\|_{L^p(\mathbb{R}^n)}, \quad 1 \leq p < n, \ \frac{1}{p^*} = \frac{1}{p} - \frac{1}{n}.$$

Proposition 3.6. *For $n \geq 3$ one has*

$$\mathcal{Y}(\mathbb{S}^n, [g_{\mathbb{S}^n}]) = c_n S_n,$$

$$S_n := \inf_{0 \not\equiv u \in C_c^\infty(\mathbb{R}^n)} \frac{\int_{\mathbb{R}^n} |\nabla u|^2 \, dx}{\|u\|^2_{L^{2^*}(\mathbb{R}^n)}}.$$

Proof. We prove the proposition in two steps:

Step 1 (Approximation). Let P_N be the north pole of $\mathbb{S}^n \subseteq \mathbb{R}^{n+1}$. Then

$$\mathcal{Y}(\mathbb{S}^n, [g_{\mathbb{S}^n}]) = \inf_{0 \not\equiv u \in C_c^\infty(\mathbb{S}^n \setminus \{P_N\})} Q_{g_{\mathbb{S}^n}}(u).$$

We already know that $\mathcal{Y}(\mathbb{S}^n, [g_{\mathbb{S}^n}]) = \inf_{u \in C^\infty(\mathbb{S}^n)} Q_{g_{\mathbb{S}^n}}(u)$. For $\varepsilon > 0$ small, consider a cutoff function $\chi_\varepsilon \in C^\infty(\mathbb{S}^n)$ such that

$$\chi_\varepsilon(x) = \begin{cases} 1 & \text{if } x \in B_\varepsilon(P_N), \\ 0 & \text{if } x \notin B_{2\varepsilon}(P_N), \end{cases}$$

and for which $|\nabla \chi_\varepsilon| \leq \frac{2}{\varepsilon}$. Then

$$\int_{\mathbb{S}^n} |\nabla \chi_\varepsilon|^2 \, dV_{g_{\mathbb{S}^n}} \leq C \int_\varepsilon^{2\varepsilon} \frac{1}{\varepsilon^2} t^{n-1} \, dt = O(\varepsilon^{n-2}).$$

This implies that for every $0 \not\equiv u \in C^\infty(\mathbb{S}^n)$ one has that $Q_{g_{\mathbb{S}^n}}((1 - \chi_\varepsilon)u) \to Q_{g_{\mathbb{S}^n}}(u)$ as $\varepsilon \to 0$.

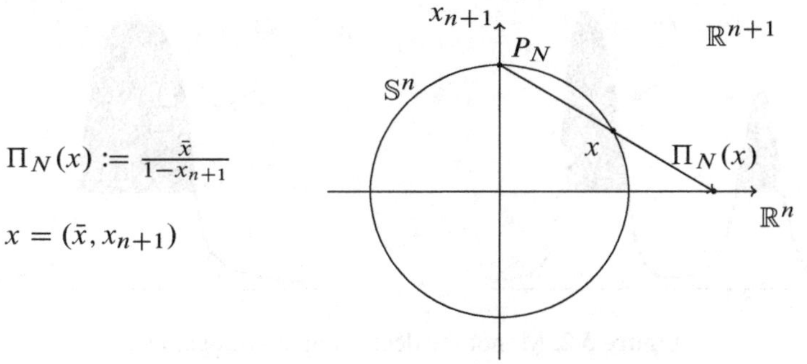

Figure 3.1. Stereographic projection

Step 2 (Conformal covariance). Consider the stereographic projection from the north pole $\Pi_N \colon \mathbb{S}^n \setminus \{P_N\} \to \mathbb{R}^n$, $\Pi_N(x) = \frac{\bar{x}}{1-x^{n+1}}$, as described in Figure 3.1. This is a conformal diffeomorphism, so there exists a function $v \colon \mathbb{S}^n \setminus \{P_N\} \to \mathbb{R}_+$ such that $\Pi_N^* g_{\mathbb{R}^n} = v^{\frac{4}{n-2}} g_{\mathbb{S}^n} =: \tilde{g}$. Using the covariance property (2.3) of L_g we write, for $u \in C_c^\infty(\mathbb{S}^n \setminus \{P_N\})$,

$$\int_{\mathbb{S}^n} \Big(c_n \Big| \nabla_{\tilde{g}} \Big(\frac{u}{v} \Big) \Big|^2 + R_{\tilde{g}} \Big| \frac{u}{v} \Big|^2 \Big) \, dV_{\tilde{g}} = \int_{\mathbb{S}^n} (c_n |\nabla_{g_{\mathbb{S}^n}} u|^2 + R_{g_{\mathbb{S}^n}} u^2) \, dV_{g_{\mathbb{S}^n}}.$$

Setting $\tilde{u} = (\frac{u}{v}) \circ \Pi_N^{-1}$, the left-hand side becomes

$$c_n \int_{\mathbb{R}^n} |\nabla \tilde{u}|^2 \, dx.$$

Similarly,

$$\|u\|_{L^{2^*}(\mathbb{S}^n)} = \|\tilde{u}\|_{L^{2^*}(\mathbb{R}^n)}.$$

Since $\tilde{u}$ is smooth and compactly supported in $\mathbb{R}^n$, the conclusion follows from the above definition of the infimum S_n. ■

The constants S_n and $\mathcal{Y}(\mathbb{S}^n[g_{\mathbb{S}^n}])$ are explicit, and derived through Proposition 3.6 and the *spherical decreasing rearrangement* procedure described below.

3.2 Spherical decreasing rearrangement

For a measurable set $A \subseteq \mathbb{R}^n$, we denote by $|A|$ the n-dimensional Lebesgue measure of A. Let $u \colon \mathbb{R}^n \to \mathbb{R}_+$ be a measurable function such that

$$\mu(t) := |\{x \colon u(x) > t\}| \quad \text{is finite for every } t > 0.$$

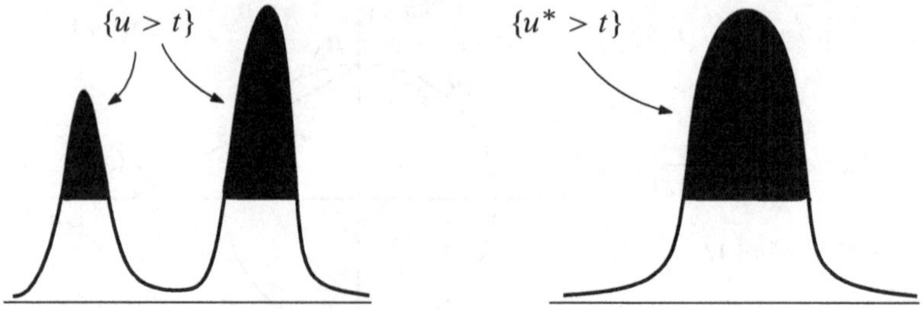

Figure 3.2. Monotone decreasing rearrangement

Then define

$$\bar{u}^*(s) := \sup\{t > 0 \colon \mu(t) > s\},$$

and set

$$u^*(x) := \bar{u}^*(\omega_n |x|), \quad \omega_n := |B_1|;$$

see Figure 3.2. Below, for $n \geq 3$ we denote by $\mathcal{D}^{1,2}(\mathbb{R}^n)$ the completion of $C_c^\infty(\mathbb{R}^n)$ with respect to the Dirichlet norm $\|\nabla \cdot\|_{L^2(\mathbb{R}^n)}$.

Proposition 3.7. *We have the following properties:*

(i) *If $u \in L^p(\mathbb{R}^n)$, then $u^* \in L^p(\mathbb{R}^n)$ and $\|u^*\|_{L^p(\mathbb{R}^n)} = \|u\|_{L^p(\mathbb{R}^n)}$.*

(ii) (Pólya–Szegö inequality) *If $u \in W^{1,\infty}(\mathbb{R}^n) \cap \mathcal{D}^{1,2}(\mathbb{R}^n)$, then $u^* \in \mathcal{D}^{1,2}(\mathbb{R}^n)$ and*

$$\|\nabla u^*\|_{L^2(\mathbb{R}^n)} \leq \|\nabla u\|_{L^2(\mathbb{R}^n)}.$$

Proof. We begin by proving (i). For $t > 0$ we let

$$E_t := \{x \colon |u(x)| > t\}$$

and

$$\chi_{E_t}(x) := \begin{cases} 1 & \text{if } |u(x)| > t, \\ 0 & \text{if } |u(x)| \leq t \end{cases} = \chi_{[0,|u(x)|)}(t).$$

By Fubini's theorem we have

$$\begin{aligned}
\int_{\mathbb{R}^n} |u(x)|^p \, dx &= \int_{\mathbb{R}^n} \int_0^\infty \chi_{[0,|u(x)|^p)}(t) \, dt \, dx = \int_{\mathbb{R}^n} \int_0^\infty \chi_{\{|u(x)|^p > t\}}(t) \, dt \, dx \\
&= \int_0^\infty |\{x \colon |u(x)| > t^{\frac{1}{p}}\}| \, dt = \int_0^\infty |\{x \colon |u^*(x)| > t^{\frac{1}{p}}\}| \, dt \\
&= \int_{\mathbb{R}^n} |u^*|^p \, dx.
\end{aligned}$$

To prove (ii) we first recall the co-area formula (see e.g. [102, 220]): for $\Omega \subseteq \mathbb{R}^n$ open and $u \in W^{1,1}(\Omega)$ one has

$$\int_\Omega |\nabla u|\, dx = \int_0^\infty P_\Omega(\{u > t\})\, dt,$$

where $P_\Omega(E)$ denotes the perimeter of E relative to Ω. If $u \in \mathcal{D}^{1,2}(\mathbb{R}^n)$ then $|E_t| < \infty$, and consequently,

$$\int_{E_t} |\nabla u|\, dx \le \|\nabla u\|_{L^2} |E_t|^{\frac{1}{2}} < \infty.$$

Consider an open set $A_t \supseteq E_t$. Applying the co-area formula with $\Omega = E_t$ we get

$$\int_{E_t} |\nabla u|\, dx = \int_0^\infty P_{A_t}(E_s)\, ds = \int_t^\infty P(E_s)\, ds.$$

From this we deduce that the function

$$t \mapsto \int_{E_t} |\nabla u|\, dx \text{ is monotone,}$$

and a.e. differentiable with

$$\frac{d}{dt}\int_{E_t} |\nabla u|\, dx = -P(E_t) \quad \text{for a.e. } t.$$

Fixing $t, h > 0$, we have

$$\frac{1}{h}\Big(\int_{E_{t+h}} |\nabla u|\, dx - \int_{E_t} |\nabla u|\, dx\Big) = -\frac{1}{h}\int_{E_t \setminus E_{t+h}} |\nabla u|\, dx,$$

$$\frac{1}{h}\int_{E_t \setminus E_{t+h}} |\nabla u|\, dx \le \Big(\frac{1}{h}\int_{E_t \setminus E_{t+h}} |\nabla u|^2\, dx\Big)^{\frac{1}{2}} \Big(\frac{1}{h}(|E_t| - |E_{t+h}|)\Big)^{\frac{1}{2}},$$

which implies that

$$-\frac{d}{dt}\int_{E_t} |\nabla u|\, dx \le -(-\mu'(t))^{\frac{1}{2}}\Big(-\frac{d}{dt}\int_{E_t} |\nabla u|^2\, dx\Big)^{\frac{1}{2}}.$$

So, at every point of differentiability we find

$$P(E_t)^2 \le |\mu'(t)|\Big(-\frac{d}{dt}\int_{E_t} |\nabla u|^2\, dx\Big).$$

Consequently,

$$\begin{aligned}\int_0^\infty \frac{P(E_t)^2}{|\mu'(t)|}\,dt &\le -\int_0^\infty \frac{d}{dt}\Big(\int_{E_t} |\nabla u|^2\,dx\Big)\,dt \\ &= \int_{E_0} |\nabla u|^2\,dx - \int_{E_\infty} |\nabla u|^2\,dx \\ &= \int_{\mathbb{R}^n} |\nabla u|^2\,dx.\end{aligned}$$

Recall the isoperimetric inequality: if $|E| < \infty$ then $|E|^{\frac{n}{n-1}} \le \kappa_n P(E)$, where

$$\kappa_n = \omega_n^{\frac{n}{n-1}} |\mathbb{S}^n|^{-1},$$

with ω_n the volume of the unit ball of $\mathbb{R}^n$. This implies

$$\int_{\mathbb{R}^n} |\nabla u|^2\,dx \ge \kappa_n^{-2} \int_0^\infty \frac{\mu(t)^{2-\frac{2}{n}}}{|\mu'(t)|}\,dt.$$

We claim that the last quantity is

$$\int_{\mathbb{R}^n} |\nabla u^*|^2\,dx.$$

In order to prove the above claim we write

$$\int_{\mathbb{R}^n} |\nabla u^*|^2\,dx = \omega_{n-1} \int_0^\infty \mu'(\tau)\tau^{n-1}\,d\tau, \quad \tau = \omega_n^{-\frac{1}{n}} \mu(t)^{\frac{1}{n}}.$$

We see that

$$\mu(u^*(\tau)) = \omega_n \tau^n \;\Rightarrow\; \mu'(u^*(\tau))(u^*)'(\tau) = n\omega_n \tau^{n-1} \;\Rightarrow\; \mu'(\tau) = n\omega_n^{\frac{1}{n}} \frac{\mu(t)^{1-\frac{1}{n}}}{\mu'(t)},$$

which, by the change of variables $\mu'(\tau)\,d\tau = dt$, leads to

$$\begin{aligned}\omega_{n-1} \int_0^\infty \mu'(\tau)\tau^{n-1}\,d\tau &= \omega_{n-1} \int_0^\infty \frac{n\omega_n^{\frac{1}{n}} \mu(t)^{1-\frac{1}{n}}}{\mu'(t)} \Big(\frac{\mu(t)}{\omega_n}\Big)^{\frac{n-1}{n}} \frac{d\tau}{dt}\,dt, \\ &= \omega_{n-1} n \omega_n^{\frac{1}{n}} \omega_n^{\frac{1-n}{n}} \int_0^\infty \frac{\mu(t)^{2-\frac{2}{n}}}{\mu'(t)}\,dt \\ &= \kappa_n^{-2} \int_0^\infty \frac{\mu(t)^{2-\frac{2}{n}}}{\mu'(t)}\,dt.\end{aligned}$$

This completes the proof. ■

Corollary 3.8. *If $u \in H^1(\mathbb{R}^n)$, then $u^* \in H^1(\mathbb{R}^n)$ and $\|u^*\|_{H^1} \le \|u\|_{H^1}$.*

3.3 The Yamabe quotient of the standard sphere

Theorem 3.9 ([14,210]). *The best constant S_n is attained within the space $\mathcal{D}^{1,2}(\mathbb{R}^n)$ by the function* $(1+|x|^2)^{\frac{2-n}{2}}$.

Proof. By Proposition 3.7 it is sufficient to study the Sobolev ratio on radial functions. Set

$$\tilde{u}(t) = e^t u(e^{at}), \quad t \in \mathbb{R},\ a := \frac{2}{n-2}.$$

By an argument similar to Step 1 in the proof of Proposition 3.6, we can also assume that u is compactly supported in $\mathbb{R}^n \setminus \{0\}$. Then $\tilde{u}$ is compactly supported in $\mathbb{R}$. Moreover, the formula

$$\tilde{u}'(t) = e^t u(e^{at}) + a e^{(a+1)t} u'(e^{at})$$

gives

$$(\tilde{u}' - \tilde{u})(t) = e^{(a+1)t} u'(e^{at}).$$

Therefore,

$$\begin{aligned}
\int_{\mathbb{R}^n} |\nabla u|^2\, dx &= |\mathbb{S}^{n-1}| \int_0^\infty r^{n-1} |u'(r)|^2\, dr \\
&= \frac{|\mathbb{S}^{n-1}|}{a} \int_{-\infty}^{\infty} (\tilde{u}' - \tilde{u})^2(t)\, dt, \qquad r = r^{at} \\
&= \frac{|\mathbb{S}^{n-1}|}{a} \int_{-\infty}^{\infty} (|\tilde{u}'(t)|^2 + |\tilde{u}(t)|^2)\, dt \\
&= \frac{|\mathbb{S}^{n-1}|}{a} \|\tilde{u}\|_{H^1(\mathbb{R}^n)},
\end{aligned}$$

where in the second-last equality we have used that $\int_{\mathbb{R}} \tilde{u}\tilde{u}'\, dt = 0$. We also have

$$\int_{\mathbb{R}^n} |u|^{2^*}\, dx = a|\mathbb{S}^{n-1}| \int_{\mathbb{R}} |\tilde{u}|^{2^*}(t)\, dt.$$

It follows that

$$S_n = \inf_{\tilde{u} \in H^1(\mathbb{R})} \frac{\frac{|\mathbb{S}^{n-1}|}{a} \|\tilde{u}\|_{H^1(\mathbb{R})}}{\left(|\mathbb{S}^{n-1}| a \|\tilde{u}\|_{L^{2^*}(\mathbb{R})}^{2^*}\right)^{\frac{2}{2^*}}}.$$

We claim that minimizers of the latter quotient exist and have the form $(\cosh at)^{-\frac{1}{a}}$, up to a translation in $\mathbb{R}$.

To prove the claim notice that we can rearrange $\tilde{u}$ in $\mathbb{R}$, so we work within the space

$$H_r^1(\mathbb{R}) := \left\{ v \in H^1(\mathbb{R}) \colon v(t) = v(-t) \text{ and } v' \le 0 \text{ on } (0,\infty) \right\}.$$

The space $H^1_r(\mathbb{R})$ embeds compactly in $L^p(\mathbb{R})$ for every $p > 2$; see e.g. [45]. Therefore, minimizers are easily seen to exist by weak lower semi-continuity of the H^1-norm.

A minimizer $v \in H^1(\mathbb{R})$ satisfies the Euler–Lagrange equation

$$-v'' + v = \mu v^{\frac{n+2}{n-2}}, \quad v > 0, \ \mu > 0. \tag{3.1}$$

Define the Hamiltonian function

$$h(t) = \frac{1}{2}v'(t)^2 - \frac{1}{2}v(t)^2 + \frac{\mu}{2}v(t)^{2^*},$$

and notice that, along solutions of (3.1), h is constant, and so $h(0) = h(t)$ for all $t \in \mathbb{R}$. Since $h \in L^1(\mathbb{R})$ (each term in the expression of h is), we get that $h \equiv 0$. Therefore, as $v'(0) = 0$, we obtain

$$\frac{1}{2}v(0)^2 = \frac{\mu}{2}v(0)^{2^*} \quad \Rightarrow \quad v(0) = \left(\frac{2^*}{2\mu}\right)^{\frac{1}{2^*-2}} =: v_0.$$

Solutions are then explicit (and unique by Cauchy's theorem) and coincide with $v = v_0(\cosh at)^{-\frac{1}{a}}$.

Translations in t for $\tilde{u}$ correspond to the transformation $u(r) \mapsto \lambda^{\frac{n-2}{2}} u(\lambda r)$, with $\lambda > 0$. Taking $\tilde{u} = v$ as above, one sees that $(1 + |x|^2)^{-\frac{n-2}{2}}$ attains S_n. ∎

Remark 3.10. We note the following:

(i) If in $\mathbb{R}^n$ we set $U(x) = (1 + |x|^2)^{\frac{2-n}{2}}$, then all other functions realizing S_n are of the form $U_{\lambda,x_0}(x) = \lambda^{\frac{n-2}{2}} U(\lambda(x - x_0))$ with $\lambda > 0$ and $x_0 \in \mathbb{R}^n$.

(ii) Extremals of the Sobolev inequality in $\mathbb{R}^n$ (via stereographic projection) correspond to solutions to the problem

$$L_{g_{\mathbb{S}^n}} u = n(n-1)u^{\frac{n+2}{n-2}} \quad \text{on } \mathbb{S}^n, u > 0. \tag{3.2}$$

These were classified by Obata in [181] and have the expression

$$\varphi_{p,\lambda} = \left(\frac{2\lambda}{\lambda^2 + 1 - (\lambda^2 - 1)\cos d_{\mathbb{S}^n}(p, x)}\right)^{\frac{n-2}{2}}, \quad p \in \mathbb{S}^n, \ \lambda > 0; \tag{3.3}$$

see Figure 3.3. They all achieve $\mathcal{Y}(\mathbb{S}^n, [g_{\mathbb{S}^n}])$, and indeed the numerator and denominator in $Q_{g_{\mathbb{S}^n}}$ are independent of p and λ. We also notice that $\varphi_{p,\lambda} \equiv 1$ for $\lambda = 1$, while as $\lambda \to \infty$ the conformal volume density $\varphi^{2^*}_{p,\lambda}$ behaves distributionally like $|\mathbb{S}^n|\delta_p$.

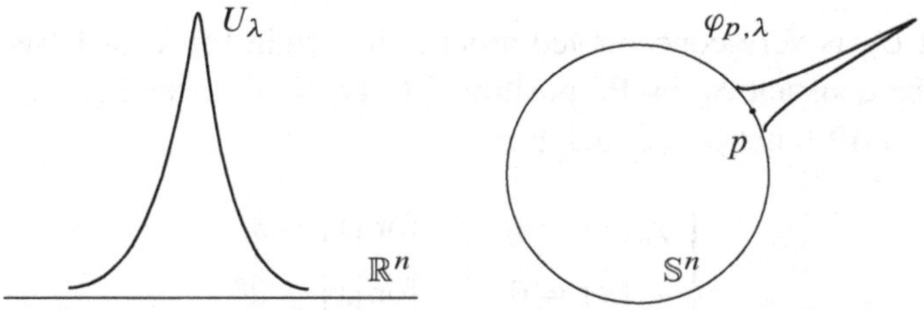

Figure 3.3. Euclidean and spherical extremal functions

(iii) The bubble functions $(\varphi_{p,\lambda})_{p,\lambda}$ form an $(n+1)$-dimensional manifold that is non-degenerate in the sense of Bott, namely the second differential of $Q_{g_{\mathbb{S}^n}}$ is (positively) non-degenerate along directions in $H^1(\mathbb{S}^n)$ orthogonal to this manifold (apart from multiplications by constants). This means that all solutions of the linearization of (3.2) at some φ_λ, namely of

$$L_{g_{\mathbb{S}^n}} v - \frac{n(n-1)(n+2)}{n-2} \varphi_{p,\lambda}^{\frac{4}{n-2}} v = 0$$

are linear combinations of derivatives of $\varphi_{p,\lambda}$ with respect to p and λ. We refer the reader to Chapter 7 for more details. Estimates related to this property, but in a quantitative form, are given in [86, 105, 106].

(iv) Working within spaces of symmetric functions, Ding [89] proved the existence of infinitely many changing-sign solutions of

$$L_{g_{\mathbb{S}^n}} u = n(n-1)|u|^{\frac{4}{n-2}} u \quad \text{on } \mathbb{S}^n,$$

corresponding to critical points of $Q_{g_{\mathbb{S}^n}}$ with larger energy. More changing-sign solutions, with specific profiles, were constructed in [83].

3.4 Existence of Yamabe metrics

We begin by recalling the following key result from [14].

Proposition 3.11. *For every closed manifold (M^n, g) one has the weak inequality $\mathcal{Y}(M,[g]) \leq \mathcal{Y}(\mathbb{S}^n,[g_{\mathbb{S}^n}])$. In the case that strict inequality holds, the infimum of Q_g is attained by a positive function yielding a constant scalar curvature metric.*

Proof. Fix a point $p \in M$, and consider normal coordinates $(x^i)_i$ centered at p. For $\lambda > 0$ large, define the function

$$U_\lambda(x) := \lambda^{\frac{n-2}{2}} U(\lambda x), \quad U(x) := \frac{1}{(1+|x|^2)^{\frac{n-2}{2}}}.$$

We note that U_λ is very concentrated around the origin for $\lambda \gg 1$ (see Figure 3.3) and attains the constant S_n by Proposition 3.6. For $\delta > 0$ small and fixed, consider now a radial cutoff function χ_δ such that

$$\begin{cases} \chi_\delta(x) = 1 & \text{for } |x| \leq \delta, \\ \chi_\delta(x) = 0 & \text{for } |x| \geq 2\delta, \\ \chi_\delta(x) \in [0,1] & \text{for all } x. \end{cases}$$

Since the metric coefficients g_{ij} converge to the Kronecker symbols as $x \to 0$, it is easy to see that

$$Q_g(\chi_\delta U_\lambda) \to \mathcal{Y}(\mathbb{S}^n, [g_{\mathbb{S}^n}]) \quad \text{as } \lambda \to \infty,$$

which implies

$$\mathcal{Y}(M^n, [g]) \leq \mathcal{Y}(\mathbb{S}^n, [g_{\mathbb{S}^n}]).$$

Suppose now that the inequality is strict. It turns out that for all $\varepsilon > 0$ there exists $C_\varepsilon > 0$ such that for every $u \in H^1(M)$ one has

$$\|u\|^2_{L^{2^*}(M)} \leq (S_n^{-1} + \varepsilon) \int_M |\nabla_g u|^2 \, dV_g + C_\varepsilon \|u\|^2_{L^2(M)}; \tag{3.4}$$

see e.g. [129, Theorem 4.6], whose proof consists in localizing the above sharp inequality in $\mathbb{R}^n$.

Suppose now $(u_j)_j$ is a minimizing sequence for Q_g: we can assume that $u_j \geq 0$ and that $\|u_j\|_{L^{2^*}}(M) = 1$ for all j. By Hölder's inequality $\|u_j\|_{L^2(M)}$ is uniformly bounded, and the fact that $(u_j)_j$ is minimizing implies that $\nabla_g u_j$ is also uniformly bounded in $L^2(M)$. Therefore, there exists $\bar{u} \in H^1(M)$ such that $u_j \rightharpoonup \bar{u}$ in $H^1(M)$. We claim that the limit is also strong in $H^1(M)$.

Assume by contradiction that this is not the case. Then it is a classical fact that

$$\begin{aligned} \int_M (c_n |\nabla_g u_j|^2 + R_g u_j^2) \, dV_g = {} & \int_M (c_n |\nabla_g \bar{u}|^2 + R_g \bar{u}^2) \, dV_g \\ & + \int_M (c_n |\nabla_g (u_j - \bar{u})|^2 + R_g (u_j - \bar{u})^2) \, dV_g \\ & + o_j(1). \end{aligned} \tag{3.5}$$

From [46] it also follows that

$$\int_M |u_j|^{2^*} \, dV_g = \int_M |\bar{u}|^{2^*} \, dV_g + \int_M |u_j - \bar{u}|^{2^*} \, dV_g + o_j(1).$$

Applying (3.4) to $u_j - \bar{u} \rightharpoonup 0$ and using the compactness of Sobolev's embedding one has that

$$\int_M (c_n|\nabla_g(u_j - \bar{u})|^2 + R_g(u_j - \bar{u})^2)\, dV_g \geq c_n S_n \|u_j - \bar{u}\|^2_{L^{2^*}(M)} + o_j(1) + o_\varepsilon(1). \tag{3.6}$$

Up to possibly passing to a subsequence, let us denote $\theta = \lim_j \|u_j - \bar{u}\|^{2^*}_{L^{2^*}(M)}$ and set

$$A = \int_M (c_n|\nabla_g \bar{u}|^2 + R_g \bar{u}^2)\, dV_g,$$
$$B = \lim_j \int_M (c_n|\nabla_g(u_j - \bar{u})|^2 + R_g(u_j - \bar{u})^2)\, dV_g.$$

We claim next that $\theta = 0$: in fact, from (3.5) we have

$$A + B = \mathcal{Y}(M, [g]),$$

which together with Proposition 3.6, (3.6), the fact that $Q_g(\bar{u}) \geq \mathcal{Y}(M, [g])$, and the arbitrariness of ε implies

$$\mathcal{Y}(M, [g]) \geq c_n S_n \theta^{\frac{2}{2^*}} + \mathcal{Y}(M, [g])(1 - \theta)^{\frac{2}{2^*}}.$$

This gives the desired claim since, by assumption, $\mathcal{Y}(M, [g]) < \mathcal{Y}(\mathbb{S}^n, [g_{\mathbb{S}^n}]) = c_n S_n$.

On the other hand, the above argument implies that $A = Q_g(\bar{u}) < \mathcal{Y}(M, [g])$ in the case $B > 0$, which would be a contradiction. Therefore, we must have $B = 0$, implying strong convergence in $H^1(M)$ of u_j to $\bar{u}$. Consequently, $\mathcal{Y}(M, [g])$ is attained by $\bar{u}$, and the Yamabe problem is solvable. Notice that $u_j \geq 0$ implies $\bar{u} \geq 0$, with strict positivity following from the maximum principle. ∎

Remark 3.12. The conclusion of Proposition 3.11 was proved in [211] under the assumption that $\mathcal{Y}(M, [g]) < \varepsilon_n$ for some small, positive, dimensional constant ε_n.

It turns out that the strict inequality $\mathcal{Y}(M, [g]) < \mathcal{Y}(\mathbb{S}^n, [g_{\mathbb{S}^n}])$ is almost always satisfied, as shown in the following result.

Theorem 3.13 ([14, 193]). *Suppose the closed manifold $(M^n, [g])$, $n \geq 3$ is not conformally equivalent to $(\mathbb{S}^n, [g_{\mathbb{S}^n}])$. Then $\mathcal{Y}(M, [g]) < \mathcal{Y}(\mathbb{S}^n, [g_{\mathbb{S}^n}])$, and therefore there exists a Yamabe metric in the conformal class $[g]$.*

We will not prove this result here, but we will just give a brief outline of the arguments involved. In view of Proposition 3.11, the main ingredient consists in finding a test function on which Q_g is lower than $\mathcal{Y}(\mathbb{S}^n, [g_{\mathbb{S}^n}])$. This was shown in complementary situations: by Aubin in dimension $n \geq 6$ for non-locally conformally flat manifolds, and by Schoen for all remaining cases.

A manifold (M, g) is said to be *locally conformally flat* if for any point p there exists a neighborhood U_p on which some conformal metric $\tilde{g} \in [g]$ is Euclidean in U_p. In dimension $n \geq 4$ this is characterized by the vanishing of the Weyl tensor; see formula (2.1).

Aubin showed in [15] the existence for any $p \in M$ of a test function $\psi_{p,\lambda}$, highly concentrated near p and close in $H^1(M)$ to a function $\varphi_{p,\lambda}$ as in Remark 3.10, such that in the limit $\lambda \to +\infty$,

$$Q_g(\psi_{p,\lambda}) = \begin{cases} \mathcal{Y}(\mathbb{S}^n, [g_{\mathbb{S}^n}]) - \tilde{c}_n |W_g|^2(p)\lambda^{-4} + o(\lambda^{-4}) & \text{for } n > 6, \\ \mathcal{Y}(\mathbb{S}^n, [g_{\mathbb{S}^n}]) - \tilde{c}_6 |W_g|^2(p)\lambda^{-4} \log \lambda + o(\lambda^{-4} \log \lambda) & \text{for } n = 6. \end{cases}$$

Here, $\tilde{c}_n$ is a positive, dimensional constant.

Schoen's construction is instead of *global type*. We saw from Proposition 3.1 that $\mathcal{Y}(M, [g])$ has the same sign as $\lambda_1(L_g)$, so if these are positive, L_g has trivial kernel and its Green's function is positive, see Remark 2.16.

Now consider a *conformal blow-up* of M at p, namely the Riemannian manifold $(M \setminus \{p\}, G_p^{\frac{4}{n-2}} g)$, where G_p stands for the Green's function of L_g with pole at p, with the convention that

$$L_g G_p = 4\frac{(n-1)^2}{n-2}\omega_{n-1}\delta_p \quad \text{distributionally on } M.$$

By the latter formula and (2.4), the scalar curvature of the metric $G_p^{\frac{4}{n-2}} g$ is identically zero on $M \setminus \{p\}$. Moreover, since G_p behaves in normal coordinates x at p like

$$G_p(x) = |x|^{2-n} + \text{l.o.t.}, \qquad x \to 0,$$

using the *inverted coordinates* $y^i = |x|^{-2} x^i$ (for $0 \neq |x|$ small enough) one can check that $(M \setminus \{p\}, G_p^{\frac{4}{n-2}} g)$ is complete and *asymptotically flat*, namely the union of a compact set and of a set which is diffeomorphic to the complement of a ball in $\mathbb{R}^n$ and on which the metric approaches the Euclidean one at infinity; see e.g. [147] and Section 10.5 for more details and precise definitions. Schoen [193] constructed test functions $\tilde{\psi}_{p,\lambda}$, also close in $H^1(M)$ to $\varphi_{p,\lambda}$, such that for λ large,

$$Q_g(\tilde{\psi}_{p,\lambda}) = \mathcal{Y}(\mathbb{S}^n, [g_{\mathbb{S}^n}]) - \hat{c}_n m(p)\lambda^{2-n} + o(\lambda^{2-n}).$$

Here, $m(p)$ is the *ADM mass* of $(M \setminus \{p\}, G_p^{\frac{4}{n-2}} g)$, as defined in [12]. By the *positive mass theorem*, proven in [198–200], $m(p) \geq 0$ for $n \leq 7$ or for (M, g) locally conformally flat; see also [201] for recent developments. Equality holds if and only if

$(M \setminus \{p\}, G_p^{\frac{4}{n-2}} g)$ is isometric to the Euclidean space, which happens precisely when the original manifold (M, g) is conformally equivalent to $(\mathbb{S}^n, g_{\mathbb{S}^n})$. Therefore, unless we are in the latter situation, one always has strict inequality in Proposition 3.11.

As proven in Theorem 4.1, the Yamabe problem admits a unique solution in negative Yamabe class, and a unique solution up to scaling in zero Yamabe class. For the positive case, uniqueness does not hold in general, as happens for the sphere or in cases discussed in [196]. Uniqueness holds, however, on Einstein manifolds different from the sphere, as a consequence of Theorem A.5 below.

Remark 3.14. The Yamabe problem is a higher-dimensional extension of the classical *uniformization problem* on closed surfaces, where one looks for conformal metrics of constant Gaussian curvature. While the original proofs relied on complex-analytic methods, the argument in [193] inspired more recently a PDE-approach that can be found in [88].

Remark 3.15. There is a special choice of conformal metric near a given point p such that geodesic normal coordinates at p for such a metric have special cancellation properties; see [147] for details. Such coordinates are called *conformal normal coordinates*, and for $n \leq 5$ or (M, g) locally conformally flat, G_p has the expression

$$G_p(x) = |x|^{2-n} + A(p) + O(|x|).$$

The map $p \mapsto A(p)$ is known as the *Robin function* of G_p, and it is proportional to $m(p)$.

While in dimension $n \geq 6$ (and (M, g) not locally conformally flat) it is a local quantity, the Weyl tensor, which determines the leading term in the correction of Q_g compared to the spherical case, in lower dimension is given by a *global quantity*, which expresses a *self-interaction* of the bubble functions with the geometry of the manifold. This difference can be heuristically explained by the fact that the bubble profiles have faster decay in larger dimension.

To give a more precise intuition for the appearance of Green's function, we notice that, at distributional level (and forgetting about dimensional constants),

$$L_g \varphi_{p,\lambda} \approx \varphi_{p,\lambda}^{\frac{n+2}{n-2}} \approx \lambda^{-\frac{n-2}{2}} \delta_p,$$

thus a proper multiple of $\lambda^{-\frac{n-2}{2}} G_p$ interpolates a bubble function with the profile of $\varphi_{p,\lambda}$ on (M, g) well, for λ large.

Notice that on $\mathbb{S}^n$ we indeed have

$$\lambda^{\frac{n-2}{2}} \varphi_{p,\lambda}(x) \to \Big(\frac{2}{1 - \cos d_{\mathbb{S}^n}(p, x)}\Big)^{\frac{n-2}{2}} = 2^{n-2} G_p(x)$$

in $C^2_{\mathrm{loc}}(\mathbb{S}^n \setminus \{p\})$.

3.5 Proof of Theorem A

With the solution of Yamabe's problem at hand, it is possible to prove Theorem A.

We first show that on every closed manifold M, given a function K negative somewhere, there exists a metric $\hat{g}$ on M with scalar curvature identically equal to K.

By Aubin's theorem in [13] there exists a metric $\bar{g}$ on M such that $R_{\bar{g}} < 0$, and hence by Proposition 3.1, $\lambda_1(L_{\bar{g}}) < 0$. Without loss of generality we can assume that $R_{\bar{g}} \equiv -1$, by Proposition 3.11. Let $p > n$, so that $W^{2,p}(M) \hookrightarrow C^1(M)$, and set

$$\Omega := \{u \in W^{2,p}(M) \colon u > 0\}.$$

Define $\Gamma \colon \Omega \times L^p(M) \to L^p(M)$ by

$$\Gamma(u, f) := L_{\bar{g}} u - f u^{\frac{n+2}{n-2}}.$$

Then Γ is of class C^1, with differential with respect to the first variable given by

$$d_u \Gamma(u, f)[v] = L_{\bar{g}} v - \frac{n+2}{n-2} f u^{\frac{4}{n-2}} v \quad \text{for all } v \in W^{2,p}(M).$$

At $(u, f) = (1, -1)$, from the fact that $R_{\bar{g}} \equiv -1$, we have that $d_u \Gamma$ is invertible. By the implicit function theorem (see e.g. [11, Chapter 2]) we can find $\varepsilon > 0$ such that if $\|f + 1\|_{L^p(M)} < \varepsilon$, then the equation $\Gamma(u, f) = 0$ has a solution u close to 1 in $W^{2,p}(M)$, so in particular u is positive on M by our choice of p.

Since K is negative somewhere, there exists $\alpha > 0$ and a smooth diffeomorphism φ such that $f = \alpha K \circ \varphi$ satisfies $\|f + 1\|_{L^p(M)} < \varepsilon$. Hence, αK is the scalar curvature of $\breve{g} = (\varphi^{-1})^* (u^{\frac{4}{n-2}} \bar{g})$, where u solves $\Gamma(u, f) = 0$. It is then sufficient to take the metric $\hat{g} = \alpha^{-1} \breve{g}$.

Suppose now that M is of type I, so there exists a metric $\mathring{g}$ with $R_{\mathring{g}} > 0$ (and hence $\lambda_1(L_{\mathring{g}}) > 0$), and that $K \geq 0$ on M.

If $K \equiv 0$, consider as before a metric $\bar{g}$ with $R_{\bar{g}} < 0$. Since $\lambda_1(L_g)$ is continuous with respect to g, interpolating between $\mathring{g}$ and $\bar{g}$ there exists $\breve{g}$ such that $\lambda_1(L_{\breve{g}}) = 0$. Still by Proposition 3.11, we can find a metric in the conformal class of $\breve{g}$ with vanishing scalar curvature.

Suppose now that $K \geq 0$, but that K is not identically zero. By the results recalled in the previous sections, there exists $g = u^{\frac{4}{n-2}} \mathring{g} \in [\mathring{g}]$ such that $R_g \equiv +1$ and such that the second differential of $Q_{\mathring{g}}(u)$ is non-negative definite. Consider again the operator

$$\Gamma(u, f) = L_g u - f u^{\frac{n+2}{n-2}}.$$

At $(u, f) = (1, 1)$ one has

$$d_u \Gamma[v] = -c_n \Delta_g v + v - \frac{n+2}{n-2} v = -\frac{4(n-1)}{n-2} \Delta_g v - \frac{4}{n-2} v.$$

Since $R_g \equiv 1$, it follows that

$$\lambda_1(d_u \Gamma(1,1)) = -\frac{4}{n-2},$$

and the above stability condition for $Q_{\mathring{g}}$ at u implies that $\lambda_2(d_u \Gamma(1,1)) \geq 0$. For generic choices of g one has $\lambda_2(d_u \Gamma(1,1)) > 0$, and therefore $d_u \Gamma(1,1)$ is invertible. We can then proceed with the same argument we used before.

Chapter 4

Variational structure of (KW-2)

In this chapter we will discuss the variational structure of the conformal curvature prescription problem (E_K). For simplicity, we will take curvatures K of the same sign as $\lambda_1(L_g)$, which coincides with that of $\mathcal{Y}(M,[g])$ by Proposition 3.1. We recall from Proposition 3.1 that for the existence of positive solutions to (4.2) with K not changing sign, the coincidence of signs is a necessary condition.

As we will see, for non-positive curvatures we will be able to find solutions as global minima of an Euler–Lagrange functional. While this property ceases to hold in positive curvature, it might be recovered on round spheres under some symmetry assumptions.

4.1 Existence and uniqueness for non-positive curvatures

Consider the following functional $I_K\colon H^1(M) \to \mathbb{R}$:

$$I_K(u) := \frac{1}{2}\int_M (c_n|\nabla_g u|_g^2 + R_g u^2)\, dV_g - \frac{1}{2^*}\int_M K|u|^{2^*}\, dV_g. \tag{4.1}$$

One notices immediately that it is well defined on $H^1(M)$ by the Sobolev embedding of this space into $L^{2^*}(M)$. It turns out that I_K is also *Fréchet differentiable* and of class C^2: we refer for these notions and properties to e.g. [11, Chapter 1].

The stationarity condition at some function $u \in H^1(M)$ means the vanishing of the differential of I_k at u, namely that $dI_K(u)[v] = 0$ for all $v \in H^1(M)$. By the results in [11] this means that

$$0 = \int_M (c_n\langle\nabla_g u, \nabla_g v\rangle + R_g uv - K|u|^{\frac{4}{n-2}}uv)\, dV_g,$$

namely that u weakly solves

$$-c_n\Delta_g u + R_g u = K|u|^{\frac{4}{n-2}}u \quad \text{on } M. \tag{4.2}$$

This equation is of the form (E_p), so Theorem 2.17 applies, providing regularity of u. If u is non-zero and non-negative, the maximum principle also yields strict positivity. We have then the following result.

Theorem 4.1. *Suppose (M^n, g), $n \geq 3$ is closed, that $\lambda_1(L_g) < 0$ (respectively, that $\lambda_1(L_g) = 0$) and that $K\colon M \to \mathbb{R}$ is negative (respectively, null). Then* (E_K) *is solvable, i.e. there exists $\tilde{g}$ conformal to g with $R_{\tilde{g}} = K$. Also, such a metric $\tilde{g}$ is unique in the negative case and unique up to a constant multiple in the null case.*

Proof. In the case that $\lambda_1(L_g) = 0$, the existence part is simply the null case in Proposition 3.1, where the conformal factor is chosen as the first eigenfunction of L_g. Uniqueness up to a constant factor follows from the simplicity of the first eigenvalue, which we discussed in Chapter 2.

Suppose now that $\lambda_1(L_g) < 0$ and that K is negative. The elementary inequality

$$R_g u^2 - \frac{1}{2^*} K |u|^{2^*} \geq \frac{1}{C} u^2 - C,$$

holding for some positive constant C, implies the following lower bound on I_K:

$$I_K(u) \geq \frac{1}{\overline{C}} \|u\|^2_{H^1} - \overline{C} \quad \text{for all } u \in H^1(M).$$

This means that I_K is *coercive*, namely $I_K(u) \to +\infty$ as $\|u\|_{H^1} \to +\infty$. By the compactness of the Sobolev embedding of $H^1(M)$ into $L^2(M)$ and Fatou's lemma, it also follows that I_K is weakly lower semi-continuous. By the reflexivity of $H^1(M)$ we can apply the *direct methods of the calculus of variations* (see [209, Chapter 1]) to find a global minimizer u of I_K.

If φ_1 is the first eigenfunction of L_g, we have that $I_K(\varepsilon\varphi_1) < I_K(0) = 0$ for ε small, and hence $u \not\equiv 0$. Since also $I_K(u) = I_K(|u|)$, we can assume that $u \geq 0$. By the above discussion, u is then a regular positive solution of (E_K), as desired.

We prove next uniqueness of $\tilde{g}$ within the given conformal class. Suppose that u_1, u_2 are two positive solutions of (E_K). We can write $u_2 = \psi u_1$ for some $\psi > 0$: then $L_g(\psi u_1) = K(\psi u_1)^{\frac{n+2}{n-2}}$, which gives

$$-c_n u_1 \Delta_g \psi - 2c_n \langle \nabla_g u_1, \nabla_g \psi \rangle_g = K\Big[(\psi u_1)^{\frac{n+2}{n-2}} - \psi u_1^{\frac{n+2}{n-2}}\Big].$$

In the above relation we used the identity

$$\Delta_g(\psi u_1) = \psi \Delta_g u_1 + 2\langle \nabla_g u_1, \nabla_g \psi \rangle_g + u_1 \Delta_g \psi,$$

and (E_K) for u_1. At a maximum point for ψ one has $\Delta_g \psi \leq 0$, $\nabla_g \psi = 0$, and hence

$$\psi^{\frac{n+2}{n-2}} - \psi \leq 0 \quad \Rightarrow \quad \psi \leq 1.$$

At a minimum point, instead

$$\Delta_g \psi \geq 0, \ \nabla_g \psi = 0 \quad \Rightarrow \quad \psi^{\frac{n+2}{n-2}} - \psi \geq 0 \quad \Rightarrow \quad \psi \geq 1.$$

Therefore, $\psi \equiv 1$ on M, and we conclude that $u_1 \equiv u_2$. ∎

Remark 4.2. If $\lambda_1(L_g) = 0$ and if one wants to prescribe a function $K \not\equiv 0$, then it cannot be $K \geq 0$ or $K \leq 0$, i.e. it has to attain both signs. This follows simply by multiplying the equation $L_g u = K u^{\frac{n+2}{n-2}}$ by φ_1 and integrating on M.

4.2 Variational approach in positive curvature

We consider now the case $\lambda_1(L_g) > 0$ and take $K > 0$ as well. Let I_K be the functional defined in formula (4.1). It turns out that in positive curvature, every non-trivial critical point is *unstable*. To see this, we notice that for any given $u \not\equiv 0$,

$$I_K(tu) = at^2 - bt^{2^*}, \quad t > 0,$$

for some $a, b > 0$. Then it follows that $\frac{d^2}{dt^2} I_K(tu)|_{t=1} < 0$. The latter formula implies also that I_K is unbounded below, and hence there is no hope of using a minimization procedure.

In analogy with the Yamabe quotient, we consider the alternative functional

$$J_K(u) := \frac{\int_M (c_n |\nabla_g u|^2 + R_g u^2)\, dV_g}{\left(\int_M K|u|^{2^*}\, dV_g\right)^{\frac{2}{2^*}}}, \quad u \in H^1(M) \setminus \{0\}. \tag{4.3}$$

This is homogeneous of degree zero and, arguing as for I_K, its non-negative critical points satisfy

$$L_g u = \mu K u^{\frac{n+2}{n-2}},$$

where μ is a Lagrange multiplier. Scaling u by a proper positive constant, one can normalize the solution so that either $\|u\|_{L^{2^*}(M)} = 1$ or $\mu = 1$, giving (E_K).

The Sobolev embedding of $H^1(M)$ into $L^{2^*}(M)$ implies that J_K is bounded below by a positive constant. One might then be tempted to look for minima of J_K in order to solve (E_K): however, it turns out that in general, even in this case, there is no minimizer, as shown by the following result.

Proposition 4.3. *For $(M^n, g) = (\mathbb{S}^n, g_{\mathbb{S}^n})$ and $K > 0$, J_K has no minimizer unless K is identically constant.*

Proof. To prove the result we notice that, clearly,

$$\inf J_K \geq \left(\max_{\mathbb{S}^n} K\right)^{-\frac{2}{2^*}} \mathcal{Y}(\mathbb{S}^n, [g_{\mathbb{S}^n}]).$$

Let $\bar{p} \in \mathbb{S}^n$ be such that $K(\bar{p}) = \max_{\mathbb{S}^n} K$. Then, if $\varphi_{p,\lambda}$ is a bubble function as in Remark 3.10, one has that

$$\varphi_{\bar{p},\lambda}^{2^*} \rightharpoonup |\mathbb{S}^n| \delta_{\bar{p}}$$

and

$$\int_{\mathbb{S}^n} \varphi_{p,\lambda} L_{g_{\mathbb{S}^n}} \varphi_{p,\lambda}\, dV_{g_{\mathbb{S}^n}} = n(n-1)|\mathbb{S}^n|, \quad \text{independently of } p \text{ and } \lambda,$$

so we get that

$$J_K(\varphi_{\bar{p},\lambda}) \xrightarrow{\lambda\to\infty} \frac{n(n-1)|\mathbb{S}^n|}{(\max_{\mathbb{S}^n} K|\mathbb{S}^{n-1}|)^{\frac{2}{2^*}}} = \inf J_K.$$

If a minimizer did exist, its support would to be contained in $\{K = K_{\max}\}$, which is impossible by the strong maximum principle unless $K \equiv K_{\max}$. ∎

It is also possible to show that the conformal volume density $|u_j|^{2^*}$ associated to a minimizing sequence $(u_j)_j$ must necessarily concentrate at a single point in the set $\{K = K_{\max}\}$; see Remark 4.7 below. One can still hope to find other critical points, of saddle type. For this aim, it is useful to introduce the following definition.

Definition 4.4. A *Palais–Smale* (or in brief, P-S) sequence for J_K is a sequence $(u_j)_j \subseteq H^1(M)$ such that, for some $c \in \mathbb{R}$,

$$J_K(u_j) \to c \in \mathbb{R} \quad \text{and} \quad \|dJ_K(u_j)\|_{H^1(M)^*} \to 0,$$

that is,

$$dJ_K(u_j)[v_j] = o_j(1)\|v_j\|_{H^1(M)} \quad \text{for every } (v_j)_j \subseteq H^1(M).$$

Remark 4.5. We notice the following properties:

(i) If $(\tilde{u}_j)_j$ is minimizing for J_K (say, normalized in $H^1(M)$), then there exists $(u_j)_j$ such that $\|u_j - \tilde{u}_j\|_{H^1(M)} \to 0$, and such that $(u_j)_j$ is a Palais–Smale sequence for J_K. This follows from Ekeland's variational principle (see [209, Chapter I]). In this case, $c = \inf J_K > 0$, because of $\lambda_1(L_g) > 0$ and of Sobolev's embedding.

(ii) Palais–Smale sequences are typically produced by min-max schemes, such as the mountain pass, by the deformation lemma (see [209, Chapter II]). If the *Palais–Smale condition* holds (i.e. P-S sequences are pre-compact), then min-max schemes generate true critical points. However, for J_K the P-S condition fails since the embedding $H^1(M) \hookrightarrow L^{2^*}(M)$ is not compact.

In the next theorem we recall the well-known profile decomposition of Palais–Smale sequences by Struwe.

Theorem 4.6 ([207] or [209]). *Suppose $\lambda_1(L_g) \geq 0$, $K > 0$, and that $(u_j)_j$ is a Palais–Smale sequence of J_K. Then, up to scaling u_j by a constant $\alpha_j > 0$, the* *following properties hold: There exist $q \in \mathbb{N} \cup \{0\}$, sequences*

$$x_j^i \to \bar{x}^i \in M, \quad \lambda_j^i \to \infty, \quad i = 1, \dots, q,$$

a solution u_0 of

$$L_g u_0 = K|u_0|^{\frac{4}{n-2}} u_0 \quad \text{on } M,$$

solutions $0 \neq u^i \in \mathcal{D}^{1,2}(\mathbb{R}^n)$ of

$$-c_n \Delta u^i = K(\bar{x}^i)|u^i|^{\frac{4}{n-2}} u^i \quad \text{on } \mathbb{R}^n, \tag{4.4}$$

such that

$$\alpha_j u_j = u_0 + \sum_{i=1}^{q} u_j^i + o_j(1) \quad \text{in } H^1(M),$$

where u_j^i are functions in $H^1(M)$ such that in normal coordinates at x_j^i,

$$(\lambda_j^i)^{-\frac{n-2}{2}} u_j^i((\lambda_j^i)^{-1} x) \xrightarrow{j\to\infty} u^i \quad \text{in } C^2_{\text{loc}}(\mathbb{R}^n).$$

Moreover, one has the splittings

$$\alpha_j^2 \int_M u_j L_g u_j \, dV_g = \int_M u_0 L_g u_0 \, dV_g + \sum_{i=1}^{q} \|u^i\|^2_{\mathcal{D}^{1,2}(\mathbb{R}^n)} + o_j(1), \tag{4.5}$$

$$|\alpha_j|^{2^*} \int_M K|u_j|^{2^*} dV_g = \int_M K|u_0|^{2^*} dV_g + \sum_{i=1}^{q} K(\bar{x}^i)\|u^i\|^{2^*}_{L^{2^*}(\mathbb{R}^n)} + o_j(1). \tag{4.6}$$

Furthermore, for all $i, \ell = 1, \dots, q$, $i \neq \ell$,

$$\frac{\lambda_j^i}{\lambda_j^\ell} + \frac{\lambda_j^\ell}{\lambda_j^i} + d_g^2(x_j^i, x_j^\ell)\lambda_j^i \lambda_j^\ell \to \infty \quad \text{as } j \to \infty. \tag{4.7}$$

Remark 4.7. The following comments are in order:

(i) If $u_j \geq 0$, then $u_0 \geq 0$ and $u^i > 0$. Positive solutions of (4.4) are classified by Theorem 5.1 below as the extremals of the Euclidean Sobolev quotient; see Theorem 3.9 and the subsequent remark.

(ii) If u_j is normalized in $H^1(M)$, or in $L^{2^*}(M)$, then there exists $C > 0$ such that $\alpha_j \in [C^{-1}, C]$.

(iii) Condition (4.7) expresses a *separation in location and scales*: to observe different limiting profiles one needs to take dilations of different orders or around points whose mutual distances are much larger than the bubble scales ($\simeq (\lambda_j^i)^{-1}$).

(iv) By the above theorem, for K positive and non-constant on $\mathbb{S}^n$, Palais–Smale sequences arising from minimizing ones (see Remark 4.5 (i)) necessarily develop a single bubble at a maximum point of K, as claimed after Proposition 4.3.

Proof of Theorem 4.6. Here we give only a general outline.

Step 1 (Subtracting the weak limit). As J_K is homogeneous of degree zero, we can assume that $(u_j)_j$ is normalized in $L^{2^*}(M)$. Since $J_K(u_j)$ is bounded, we get that $(u_j)_j$ is bounded in $H^1(M)$, and hence $\alpha_j u_j \rightharpoonup u_0$, where u_0 solves (4.2), and $v_j := u_j - u_0$ is also a Palais–Smale sequence.

Step 2 (Concentration-compactness and extraction of one bubble). For $r > 0$ small, consider the quantity

$$F_j(r) := \sup_{x\in M} \int_{B_r(x)} |\nabla u_j|^2 \, dV_g.$$

From its definition, it follows that F_j is monotone in r. Given $\varepsilon > 0$, let r_j, x_j be such that $F_j(r_j) = \varepsilon$, the supremum is attained at x_j, and r_j is the *smallest scale* at which energy concentrates. It is possible to show that, for ε small enough, in normal coordinates x at x_j,

$$w_j(x) := r_j^{\frac{n-2}{2}} v_j(r_j x)$$

converges in $\mathcal{D}^{1,2}_{\mathrm{loc}}(\mathbb{R}^n)$ to a solution $\overline{w}$ of

$$-\Delta \overline{w} = K(\bar{x})|\overline{w}|^{\frac{4}{n-2}} \overline{w} \quad \text{on } \mathbb{R}^n,$$

for $x_j \to \bar{x}$. By Sobolev's inequality, there exists $C > 0$ such that

$$\|\overline{w}\|_{\mathcal{D}^{1,2}(\mathbb{R}^n)} \geq \frac{1}{C}.$$

Step 3 (Iteration). Now consider, up to using a proper cutoff function so that u^1 is well defined on M and normal coordinates x are still centered at x_j,

$$\tilde{v}_j := v_j - (\lambda_j^1)^{\frac{n-2}{2}} u^1(\lambda_j^1 x), \quad \lambda_j^1 := r_j^{-1}, \; u^1 := \overline{w}.$$

Then $(\tilde{v}_j)$ is again a Palais–Smale sequence for J_K, and we can repeat the above procedure. The iteration stops after finitely many times since the initial norm is controlled and we are removing a quantity positively bounded from below at each step.

The splittings (4.5) and (4.6) are in the same spirit of some arguments in the proof of Proposition 3.11. ■

Remark 4.8. In the case of a positive and constant curvature function K, Theorem 4.6 includes Proposition 3.11 as a particular case.

4.3 Proof of Theorem B

As a second application of Theorem 4.6, in addition to the one in the last remark, we prove Theorem B from the introduction, which is a particular case of the main result in [100]. We can assume without loss of generality that $\max_{\mathbb{S}^3} K = 1$, since if we scale K by a constant, the same would happen to the solutions of (E_K). Let J_K be the functional defined in (4.3). We then have

$$\inf_{H^1(\mathbb{S}^3)} J_K = \mathcal{Y}(\mathbb{S}^3, [g_{\mathbb{S}^3}]),$$

and that the infimum is not attained, by Proposition 4.3.

However, exploiting the symmetry of K, we can restrict ourselves to the subspace

$$H_s^1(\mathbb{S}^3) := \{u \in H^1(\mathbb{S}^3) : u(x) = u(-x)\}.$$

Applying Theorem 4.6 to Palais–Smale sequences in $H_s^1(\mathbb{S}^3)$, one can see that the minimal energy of a non-compact sequence occurs where there is formation of two bubbles at antipodal points. To compute this value, define (see Figure 4.1)

$$\Phi_{p,\lambda}(x) := \varphi_{p,\lambda}(x) + \varphi_{-p,\lambda}(x).$$

Notice that

$$\int_{\mathbb{S}^3} \Phi_{p,\lambda} L_{g_{\mathbb{S}^3}} \Phi_{p,\lambda} \, dV_{g_{\mathbb{S}^3}} = 2 \int_{\mathbb{S}^3} \varphi_{p,\lambda} L_{g_{\mathbb{S}^3}} \varphi_{p,\lambda} \, dV_{g_{\mathbb{S}^3}} + o_\lambda(1),$$

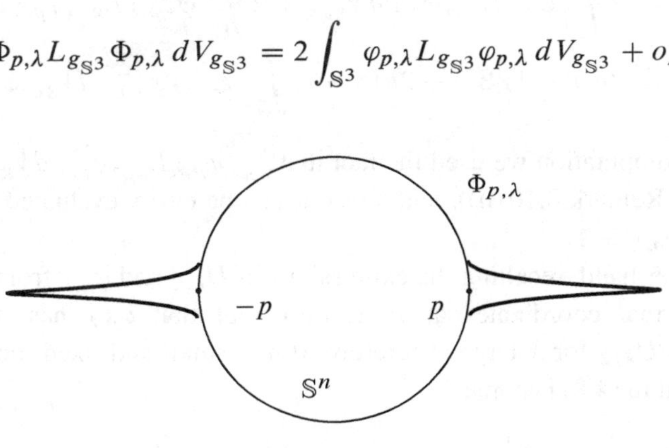

Figure 4.1. The double bubble $\Phi_{p,\lambda}$

and that

$$\int_{\mathbb{S}^3} \Phi_{p,\lambda}^{2^*} \, dV_{g_{\mathbb{S}^3}} = 2 \int_{\mathbb{S}^3} \varphi_{p,\lambda}^{2^*} \, dV_{g_{\mathbb{S}^3}} + o_\lambda(1).$$

If we then choose p so that $K(p) = \max_{\mathbb{S}^3} K = 1$, we have

$$J_K(\Phi_{p,\lambda}) \to 2^{1-\frac{2}{2^*}} \mathcal{Y}(\mathbb{S}^3, [g_{\mathbb{S}^3}]) > \mathcal{Y}(\mathbb{S}^3, [g_{\mathbb{S}^3}]).$$

By the above observation, if there exists $u \in H_s^1(\mathbb{S}^3)$ with $J_K(u) < 2^{1-\frac{2}{2^*}} \mathcal{Y}(\mathbb{S}^3, [g_{\mathbb{S}^3}])$, then J_K attains a minimum on $H_s^1(\mathbb{S}^3)$. In fact, by Remark 4.5 we obtain a Palais–Smale sequence in $H_s^1(\mathbb{S}^3)$ at a level below $2^{1-\frac{2}{2^*}} \mathcal{Y}(\mathbb{S}^3, [g_{\mathbb{S}^3}])$, which strongly converges by Theorem 4.6. Then applying *Palais' symmetric criticality principle* from [184], we have the vanishing of Lagrange's multiplier from the constraint to $H_s^1(\mathbb{S}^3)$, and then get a solution of (E_K).

We see in the next proposition that, under the assumptions of Theorem B, the strict inequality indeed holds true.

Proposition 4.9. *Let p be a maximum point of K and assume that $K(p) = 1$. Then for λ large one has*

$$J_K(\Phi_{p,\lambda}) < 2^{1-\frac{2}{2^*}} \mathcal{Y}(\mathbb{S}^3, [g_{\mathbb{S}^3}]),$$

hence a minimum of J_K in $H_s^1(\mathbb{S}^3)$ exists.

Proof. To highlight the role of the dimension in this proof, we carry out the expansions for general n, specializing only at the end to the three-dimensional case.

Using simple calculations, we find that the numerator in $J_K(\Phi_{p,\lambda})$ is

$$\begin{aligned} &\int_{\mathbb{S}^3} \Phi_{p,\lambda} L_{g_{\mathbb{S}^3}} \Phi_{p,\lambda} \, dV_{g_{\mathbb{S}^3}} \\ &\quad = 2 \int_{\mathbb{S}^3} \varphi_{p,\lambda} L_{g_{\mathbb{S}^3}} \varphi_{p,\lambda} \, dV_{g_{\mathbb{S}^3}} + 2 \int_{\mathbb{S}^3} \varphi_{-p,\lambda} L_{g_{\mathbb{S}^3}} \varphi_{p,\lambda} \, dV_{g_{\mathbb{S}^3}} \\ &\quad \overset{n=3}{=} 2n(n-1)|\mathbb{S}^n| + 2n(n-1) \int_{\mathbb{S}^3} \varphi_{-p,\lambda} \varphi_{p,\lambda}^{\frac{n+2}{n-2}} \, dV_{g_{\mathbb{S}^3}}. \end{aligned} \tag{4.8}$$

In the above computation we used the fact that $\int_{\mathbb{S}^3} \varphi_{p,\lambda} L_{g_{\mathbb{S}^3}} \varphi_{p,\lambda} \, dV_{g_{\mathbb{S}^3}}$ is independent of λ (see Remark 3.10 (ii)), and hence it can be easily evaluated taking $\lambda = 1$, that is, when $\varphi_{p,\lambda} \equiv 1$.

On the other hand, recalling the expressions of $U_{\lambda,x}$ and $\varphi_{p,\lambda}$ from Chapter 3, in conformal normal coordinates at p, it turns out that $\varphi_{p,\lambda}$ has the expression $2^{\frac{n-2}{2}} U_{\lambda/2,0} = U_{\lambda/2}$ for λ large. Therefore, if δ is small and fixed, one can see that the last integral in (4.8) becomes

$$\varphi_{-p,\lambda}(p) 2^{\frac{n+2}{2}} \int_{B_\delta(0)} U_{\lambda/2}^{\frac{n+2}{n-2}} \, dx + \varphi_{-p,\lambda}(p)^{\frac{n+2}{n-2}} 2^{\frac{n-2}{2}} \int_{B_\delta(0)} U_{\lambda/2} \, dx + O(\lambda^{-n}).$$

After using some changes of variables and the above formulas, we then get

$$\int_{\mathbb{S}^3} \Phi_{p,\lambda} L_{g_{\mathbb{S}^3}} \Phi_{p,\lambda}\, dV_{g_{\mathbb{S}^3}} = 2n(n-1)|\mathbb{S}^n| + c_0\lambda^{2-n} + o(\lambda^{2-n}),$$

where $c_0 = n(n-1)2^{n+1} \int_{\mathbb{R}^n} U(x)^{\frac{n+2}{n-2}}\, dx$.

To estimate the denominator in $J_K(\Phi_{p,\lambda})$ we use a Taylor expansion of $\Phi_{p,\lambda}$ around $\varphi_{p,\lambda}$ near p and around $\varphi_{-p,\lambda}$ near $-p$ to find

$$\int_{\mathbb{S}^3} K\Phi_{p,\lambda}^{2^*}\, dV_{g_{\mathbb{S}^3}} = 2\int_{\mathbb{S}^3} K\varphi_{p,\lambda}^{2^*}\, dV_{g_{\mathbb{S}^3}} + 22^* \int_{\mathbb{S}^3} K\varphi_{p,\lambda}^{\frac{n+2}{n-2}} \varphi_{-p,\lambda}\, dV_{g_{\mathbb{S}^3}} + o(\lambda^{2-n}),$$

and therefore, recalling that $K(p) = 1$, an expansion similar to the one for the numerator gives

$$\int_{\mathbb{S}^3} K\Phi_{p,\lambda}^{2^*}\, dV_{g_{\mathbb{S}^3}} = 2\int_{\mathbb{S}^3} K\varphi_{p,\lambda}^{2^*}\, dV_{g_{\mathbb{S}^3}} + 2^* c_0\lambda^{2-n} + o(\lambda^{2-n}).$$

Using conformal covariance we find that

$$\int_{\mathbb{S}^3} K\varphi_{p,\lambda}^{2^*}\, dV_{g_{\mathbb{S}^3}} = \int_{\mathbb{R}^n} KU_{\lambda/2}^{2^*}\, dx,$$

where in the latter integral, with an abuse of notation, we are composing K with the inverse stereographic projection from $-p$. Using a change of variables and a Taylor expansion of K, whose first-order terms vanish due to the fact that p is a maximum point, and recalling that

$$U_{\lambda/2}(x) = (\lambda/2)^{\frac{n-2}{2}} U\Big(\frac{\lambda}{2}x\Big), \quad U(x) = (1+|x|^2)^{-\frac{n-2}{2}},$$

we obtain

$$\begin{aligned}
\int_{\mathbb{R}^n} KU_\lambda^{2^*}\, dx &= \int_{\mathbb{R}^n} K(2\lambda^{-1}y)U^{2^*}(y)\, dy \\
&= \int_{\mathbb{R}^n} \Big(K(0) + 2\lambda^{-2} \sum_{i,j=1}^{n} K_{x^i x^j}(0) y^i y^j\Big) U^{2^*}(y)\, dy + o(\lambda^{-2}) \\
&= K(0)\int_{\mathbb{R}^n} U^{2^*}\, dy + 2\lambda^{-2} \sum_{i=1}^{n} K_{x^i x^i}(0) \int_{\mathbb{R}^n} (y^i)^2 U^{2^*}\, dy \\
&\quad + o(\lambda^{-2}) \\
&= |\mathbb{S}^n| K(0) + c_1\lambda^{-2}\Delta K(0) + o(\lambda^{-2}),
\end{aligned}$$

where we have set $c_1 = \frac{2}{n}\int_{\mathbb{R}^n} |y|^2 U(y)^{2^*}\, dy$ and also used the fact that $|y|^2 = \sum_{i=1}^n (y^i)^2$. This gives

$$\int_{\mathbb{S}^3} K\Phi_{p,\lambda}^{2^*}\, dV_{g_{\mathbb{S}^3}} = 2\,\underbrace{K(p)}_{=1}\,|\mathbb{S}^n| + 2^* c_0 \lambda^{2-n} + 2c_1\lambda^{-2}\Delta K(p) + o(\lambda^{-2}). \tag{4.9}$$

Therefore, since $n = 3$ one finds that

$$\begin{aligned} J_K(\Phi_{p,\lambda}) &= \frac{2n(n-1)|\mathbb{S}^n| + c_0\lambda^{2-n} + o(\lambda^{2-n})}{(2|\mathbb{S}^n| + 2^* c_0 \lambda^{2-n} + o(\lambda^{2-n}) + O(\lambda^{-2}))^{\frac{2}{2^*}}} \\ &\overset{n=3}{=} 2^{1-\frac{2}{2^*}} \mathcal{Y}(\mathbb{S}^3, [g_{\mathbb{S}^3}]) - c_2\lambda^{-1} + o(\lambda^{-1}), \end{aligned} \tag{4.10}$$

for some positive constant c_2. This yields the desired result. ∎

Remark 4.10. We have the following observations:

(i) In Theorem B, antipodal symmetry can be replaced by the action of a finite group Γ of isometries without fixed points. In general dimension n, the same proof works if at its maximum points K is *flat* of order $\beta > n - 2$, namely if in the Taylor expansion of K no homogeneous term of order less than or equal to $n - 2$ appears. In this case, for the expansion in (4.9) the term $2^* c_0 \lambda^{2-n}$ is still leading among the infinitesimal ones.

(ii) Recalling Remark 3.15, we have that in $C^2_{\mathrm{loc}}(\mathbb{S}^3 \setminus \{p, -p\})$,

$$\lambda^{\frac{1}{2}} \Phi_{p,\lambda} \to 2G_p^{L_{g_{\mathbb{S}^3}}} + 2G_{-p}^{L_{g_{\mathbb{S}^3}}} = 2G_p^{L_{g_{\mathbb{RP}^3}}}.$$

Such a *multi-pole* Green's function on $\mathbb{S}^3$ coincides after passing to the quotient with the Green's function of the conformal Laplacian on $\mathbb{S}^3/\Gamma$, where Γ stands for the antipodal group.

(iii) Other results imposing symmetries, possibly with fixed points, can be found in [8, 128, 132, 214].

(iv) The result in [178] concerning prescription of antipodally symmetric functions on $\mathbb{S}^2$ was also proved via a minimization procedure within a space of symmetric functions, reducing the problem to the two-dimensional projective space.

Chapter 5

Classification of entire solutions in the Euclidean space

We proved in Section 3.3 that the best constant

$$S_n = \inf_{0 \neq u \in \mathcal{D}^{1,2}} \frac{\int_{\mathbb{R}^n} |\nabla u|^2 \, dx}{\|u\|^2_{L^{2^*}(\mathbb{R}^n)}}, \qquad n \geq 3$$

is realized by $U(x) = (1 + |x|^2)^{-\frac{n-2}{2}}$, the functions U_{λ,x_0} in Remark 3.10, and their constant multiples. Such extremals satisfy

$$-\Delta u = \mu u^{\frac{n+2}{n-2}} \quad \text{in } \mathbb{R}^n,$$

where μ is a positive Lagrange multiplier. Specifically, for U as above one has $\mu = n(n-2)$. The theorem below, proved by Caffarelli–Gidas–Spruck, indeed classifies *all* positive solutions of the above equation in the whole of Euclidean space.

Theorem 5.1 ([47]). *Suppose $u \in C^2(\mathbb{R}^n)$ solves*

$$-\Delta u = n(n-2) u^{\frac{n+2}{n-2}} \quad \textit{on } \mathbb{R}^n, \ u > 0. \tag{E}$$

Then there exist $\lambda > 0$ and $\bar{x} \in \mathbb{R}^n$ such that $u = U_{\lambda,\bar{x}}$.

Remark 5.2. In the above theorem there are neither decay nor integrability assumptions on u, and therefore it is useful to perform *blow-up analysis* of solutions to (E_K) and ($E_{K,\tau}$), which will be the goal of the next chapter. A proof of Theorem 5.1 under certain decay assumptions at infinity was given in [111]; see also [181].

5.1 Moving planes and moving spheres

The original proof used a *moving planes* technique, consisting in comparing a solution to its reflection across a plane and using the maximum principle (see [4, 111]). Here we present an argument due to Li and Zhang [156], based on the variant of *moving spheres*, exploiting *Kelvin's inversion* (see also [73]).

The Kelvin inversion is defined as the map sending $0 \neq x \in \mathbb{R}^n$ to $\frac{x}{|x|^2}$: it preserves the sphere $\mathbb{S}^{n-1}$ and it is a conformal map. In fact, if Π_N is the stereographic projection as defined in the proof of Proposition 3.6, and if $\mathcal{T}$ stands for the reflection on $\mathbb{S}^n$ with respect to the plane $\{y^{n+1} = 0\}$, then the Kelvin inversion can be obtained as

$$\Pi_N \circ \mathcal{T} \circ \Pi_N^{-1}.$$

For $z \in \mathbb{R}^n$ and $\lambda > 0$, define

$$u_{z,\lambda} := \frac{\lambda^{n-2}}{|x-z|^{n-2}} u\Big(z + \frac{\lambda^2(x-z)}{|x-z|^2}\Big), \quad x \neq z.$$

If u solves (E), then $u_{z,\lambda}$ solves the same equation on $\mathbb{R}^n \setminus \{z\}$. Notice that

$$x \mapsto z + \frac{\lambda^2(x-z)}{|x-z|^2}$$

is an inversion with respect to $\partial B_\lambda(z)$. Fixing $z \in \mathbb{R}^n$, for $\lambda > 0$ small we have $u_{z,\lambda} < u$ in an exterior neighborhood of $\partial B_\lambda(z)$. Next we prove a more global comparison property.

Lemma 5.3. *For every $z \in \mathbb{R}^n$ there exists $\lambda_0(z)$ such that*

$$u_{z,\lambda}(x) \leq u(x) \quad \textit{for every } 0 < \lambda < \lambda_0(z),\ x \notin B_\lambda(z).$$

Proof. Without loss of generality we assume that $z = 0$, and set $u_\lambda = u_{0,\lambda}$. Working in polar coordinates, there exists $r_0 > 0$ such that

$$\frac{d}{dr}(r^{\frac{n-2}{2}} u(r,\theta)) > 0 \quad \text{for } r \in (0, r_0).$$

Since $r^{\frac{n-2}{2}} u_\lambda(r,\theta) = (\lambda^2/r)^{\frac{n-2}{2}} u(\lambda^2/r, \theta)$, this implies that

$$u_\lambda(x) < u(x) \quad \text{for } |x| \in (\lambda, r_0).$$

As u is super-harmonic, there holds

$$u(x) \geq \big(\min_{\partial B_{r_0}} u\big) \frac{r_0^{n-2}}{|x|^{n-2}} \quad \text{for } |x| \geq r_0.$$

Setting

$$\lambda_0 := r_0 \Big(\frac{\min_{\partial B_{r_0}} u}{\max_{\partial B_{r_0}} u}\Big)^{\frac{1}{n-2}} \leq r_0,$$

we have that for $\lambda \in (0, \lambda_0)$,

$$u_\lambda(x) \leq \frac{\lambda_0^{n-2}}{|x|^{n-2}} \big(\max_{\bar{B}_{r_0}} u\big) \leq \frac{r_0^{n-2} \min_{\partial B_{r_0}} u}{|x|^{n-2}} \quad \text{for } |x| \geq r_0.$$

Combining the above estimates one gets

$$u_\lambda(x) \leq u(x) \quad \text{for } |x| \geq \lambda,$$

as desired. ∎

Lemma 5.4. *Define*

$$\bar{\lambda}(z) := \sup\{\mu > 0 : u_{z,\lambda}(x) \leq u(x)\ \forall 0 < \lambda \leq \mu,\ |x - z| \geq \lambda\}.$$

If $\bar{\lambda}(z) < \infty$ *for some* $z \in \mathbb{R}^n$, *then* $u_{z,\bar{\lambda}(z)} \equiv u$ *in* $\mathbb{R}^n \setminus \{z\}$.

Proof. Again, without loss of generality we assume that $z = 0$, set $\bar{\lambda} = \bar{\lambda}(0)$, $u_{\bar{\lambda}} = u_{0,\bar{\lambda}}$, and

$$\Sigma_\lambda := \{y : |y| > \lambda\}.$$

It is sufficient to show that $u = u_{\bar{\lambda}}$ on $\Sigma_{\bar{\lambda}}$ since (either by unique continuation, or by the fact that if $u \approx \frac{1}{|x|^{n-2}}$ then one can use the result in [111]), this would imply $u \equiv u_{\bar{\lambda}}$ on $\mathbb{R}^n \setminus \{0\}$.

Clearly, we have that $u_{\bar{\lambda}} \leq u$ on $\Sigma_{\bar{\lambda}}$: this yields

$$-\Delta(u - u_{\bar{\lambda}}) = n(n-2)\big(u^{\frac{n+2}{n-2}} - u_{\bar{\lambda}}^{\frac{N+2}{n-2}}\big) \geq 0 \quad \text{on } \Sigma_{\bar{\lambda}}.$$

If $u = u_{\bar{\lambda}}$ on $\Sigma_{\bar{\lambda}}$, we are done. Otherwise, we can use Hopf's lemma on $\partial B_{\bar{\lambda}}$ to find

$$\frac{d}{dr}(u - u_{\bar{\lambda}}) \geq b > 0 \quad \text{on } \partial B_{\bar{\lambda}},$$

for some $b > 0$. By continuity of ∇u, there exists $R > \bar{\lambda}$ such that

$$\frac{d}{dr}(u - u_\lambda) \geq \frac{b}{2} \quad \text{for } \lambda \in [\bar{\lambda}, R], r \in [\lambda, R].$$

Since $u = u_\lambda$ on ∂B_λ, we have

$$u(x) - u_\lambda(x) > 0 \quad \text{for } \lambda \in [\bar{\lambda}, R], |x| \in (\lambda, R].$$

Setting

$$c = \min_{\partial B_R}(u - u_{\bar{\lambda}}),$$

by the fact that $u - u_{\bar{\lambda}}$ is super-harmonic, it follows that

$$u(x) - u_{\bar{\lambda}}(x) \geq \frac{cR^{n-2}}{|x|^{n-2}}, \quad |x| \geq R.$$

This implies

$$u(x) - u_\lambda(x) \geq \frac{cR^{n-2}}{|x|^{n-2}} - (u(x) - u_{\bar{\lambda}}(x)), \quad |x| \geq R.$$

By the uniform continuity of u on $\bar{B}_R$ there exists $\varepsilon \in (0, R - \bar{\lambda})$ such that

$$\Big|\lambda^{n-2}u\Big(\frac{\lambda^2 x}{|x|^2}\Big) - \bar{\lambda}^{n-2}u\Big(\frac{\bar{\lambda}^2 x}{|x|^2}\Big)\Big| \leq c\frac{R^{n-2}}{2} \quad \text{for } \lambda \in [\bar{\lambda}, \bar{\lambda} + \varepsilon],\ |x| \geq R.$$

Hence

$$u(x) - u_\lambda(x) > 0 \quad \text{for } \lambda \in [\bar{\lambda}, \bar{\lambda} + \varepsilon],\ |x| \geq R,$$

a contradiction to the definition of $\bar{\lambda}$. ■

Lemma 5.5. *If $\bar{\lambda}(\bar{z}) = \infty$ for some $\bar{z} \in \mathbb{R}^n$, then $\bar{\lambda}(z) = \infty$ for every $z \in \mathbb{R}^n$.*

Proof. Since $\bar{\lambda}(\bar{z}) = \infty$, we have

$$u(x) \geq u_{\bar{z},\lambda}(x) \quad \text{for every } \lambda > 0,\ |x - z| \geq \lambda,$$

which implies that

$$\lim_{|x|\to\infty} |x|^{n-2}u(x) = \infty.$$

If instead $\bar{\lambda}(z) < \infty$ for some $z \in \mathbb{R}^n$, by the previous lemma

$$\lim_{|x|\to\infty} |x|^{n-2}u(x) = \lim_{|x|\to\infty} |x|^{n-2}u_{z,\bar{\lambda}(z)}(x) = \bar{\lambda}(z)^{n-2}u(z) < \infty,$$

which is a contradiction. ■

Lemma 5.6. *We have that $\bar{\lambda}(z) < \infty$ for every $z \in \mathbb{R}^n$.*

Proof. If by contradiction $\bar{\lambda}(\bar{z}) = \infty$ for some $\bar{z} \in \mathbb{R}^n$, then by Lemma 5.5, $\bar{\lambda}(z) = \infty$ for every $z \in \mathbb{R}^n$, that is

$$u_{z,\lambda}(x) \leq u(x) \quad \text{for every } \lambda > 0,\ |x - z| \geq \lambda.$$

We claim that the above relation would imply that u is identically constant, which is impossible since it is positive and satisfies (E).

In order to prove the claim we define

$$w_{z,\lambda}(x) = u(z + x) - \Big(\frac{\lambda}{|x|}\Big)^{n-2} u\Big(z + \frac{\lambda^2 x}{|x|^2}\Big), \quad |x| \geq \lambda.$$

Then

$$w_{z,|x|}(x) = 0 \quad \text{and} \quad w_{z,|x|}(rx) \geq 0 \ \text{ for } r \geq 1,$$

which implies

$$\frac{d}{dr}(w_{z,|x|}(rx))|_{r=1} \geq 0.$$

Notice that

$$w_{z,|x|}(x) = u(z + rx) - r^{2-n} u\Big(z + \frac{x}{r}\Big),$$

and hence

$$\begin{aligned} 0 \le \frac{d}{dr}(w_{z,|x|}(rx))|_{r=1} &= \nabla u(z + x) \cdot x + (n - 2)u(z + x) + \nabla u(z + x) \cdot x \\ &= 2\nabla u(z + x) \cdot x + (n - 2)u(z + x). \end{aligned}$$

Changing variables, this becomes

$$2\nabla u(y) \cdot (y - z) + (n - 2)u(y) \ge 0 \quad \text{for every } y, z \in \mathbb{R}^n.$$

Dividing by $|z|$ and sending $|z|$ to $+\infty$ one finds

$$\nabla u(y) \cdot \theta = 0 \quad \text{for every } y \in \mathbb{R}^n,\ \theta \in \mathbb{S}^{n-1}.$$

This implies the constancy of u, as claimed. ■

Proof of Theorem 5.1. By Lemmas 5.4 and 5.6, for every $z \in \mathbb{R}^n$ there exists $\lambda > 0$ such that $u_{z,\bar{\lambda}(z)} \equiv u$ on $\mathbb{R}^n \setminus \{0\}$, that is,

$$u(y) = \Big(\frac{\bar{\lambda}(z)}{|y - z|}\Big)^{n-2} u\Big(z + \frac{\bar{\lambda}(z)^2 (y - z)}{|y - z|^2}\Big). \tag{5.1}$$

This implies that

$$B := \lim_{|y|\to\infty} |y|^{n-2} u(y) = \bar{\lambda}(z)^{n-2} u(z) \quad \text{for every } z \in \mathbb{R}^n.$$

If $B = 0$ then $u = 0$, which is impossible. Therefore, without loss of generality we can assume that $B = 1$. Taylor-expanding (5.1) at $y = 0$ and at $y = z$, we find

$$\begin{aligned} u(y) &= \Big(\frac{\bar{\lambda}(0)}{|y|}\Big)^{n-2}\Big(u(0) + \frac{\bar{\lambda}(0)^2}{|y|^2}\nabla u(0) \cdot y + o\Big(\frac{1}{|y|}\Big)\Big) && \text{as } |y| \to \infty, \\ u(y) &= \Big(\frac{\bar{\lambda}(z)}{|y - z|}\Big)^{n-2}\Big(u(z) + \frac{\bar{\lambda}(z)^2}{|y - z|^2}\nabla u(z) \cdot (y - z) + o\Big(\frac{1}{|y|}\Big)\Big) && \text{as } |y| \to \infty. \end{aligned}$$

Combining the facts that $u_{z,\bar{\lambda}(z)} = u$ and $B = 1$, we can write the above relations as

$$\frac{1}{|y|^{n-2}} + \frac{y \cdot \nabla u(0)}{u(0)^{\frac{n}{n-2}} |y|^n} = \frac{1}{|y - z|^{n-2}} + \frac{(y - z) \cdot \nabla u(z)}{u(z)^{\frac{n}{n-2}} |y - z|^n} + \text{l.o.t.}$$

Again, by a Taylor expansion,

$$\frac{1}{|y - z|^{n-2}} = \frac{1}{|y|^{n-2}} + (n - 2)\frac{y \cdot z}{|y|^n} + \text{l.o.t.} \quad \text{as } |y| \to \infty,$$

and hence

$$u(0)^{-\frac{n}{n-2}} y \cdot \nabla u(0) = (n-2) y \cdot z + u(z)^{-\frac{n}{n-2}} y \cdot \nabla u(z),$$

from which we obtain, for every $z \in \mathbb{R}^n$,

$$u(z)^{-\frac{n}{n-2}} \nabla u(z) = u(0)^{-\frac{n}{n-2}} \nabla u(0) - (n-2) z.$$

Since $\nabla u^{\frac{-2}{n-2}}(z) = \frac{-2}{n-2} u(z)^{-\frac{n}{n-2}} \nabla u(z)$, the above formula implies that

$$u(z)^{-\frac{2}{n-2}} = |z - \bar{z}|^2 + d,$$

for some $\bar{z} \in \mathbb{R}^n$ and some $d > 0$. ∎

Remark 5.7. The following facts hold true:

(i) In $\mathbb{R}^n$, positive solutions of the equation $\Delta u = 0$ are identically constants by the classical Liouville theorem, and there do not exist positive solutions of $\Delta u = u^q$, $q > 1$. This can be shown for example by considering the radial average $\bar{u}(r)$ of u on a sphere of radius r and centered at the origin. Using the fact that the Laplacian commutes with the radial average together with Jensen's inequality we find that $\bar{u}$ would solve

$$\Delta \bar{u} \geq \bar{u}^q.$$

By classical ODE theory, the solution of such a differential inequality would blow up in finite time, contradicting the fact that u is entire.

(ii) On $\mathbb{R}^2$, a counterpart of Theorem 5.1 was proved in [67]. The latter regards entire solutions of the equation

$$-\Delta u = e^{2u} \quad \text{in } \mathbb{R}^2.$$

Assuming the right-hand side of the equation is integrable on $\mathbb{R}^2$, the authors still classified solutions as conformal factors of the stereographic projection. Removing the integrability assumption, the structure of solutions can instead become more rich; see e.g. [3].

5.2 Liouville theorem in the sub-critical case

In this section we prove a rigidity result by Gidas–Spruck for entire solutions of sub-critical equations.

Theorem 5.8 ([112]). *Let $u \in C^2(\mathbb{R}^n)$ be a non-negative solution of*

$$-\Delta u = u^q \ \text{in} \ \mathbb{R}^n, \quad 1 < q < \frac{n+2}{n-2}.$$

Then $u \equiv 0$.

Here we present a proof due to Chen–Li [67] based on the moving planes technique. We set

$$v(x) := \frac{1}{|x|^{n-2}} u\Big(\frac{x}{|x|^2}\Big).$$

Then v satisfies

$$-\Delta v = \frac{1}{|x|^\alpha} v^q \ \text{on} \ \mathbb{R}^n \setminus \{0\}, \quad \alpha := n + 2 - q(n-2) > 0. \tag{5.2}$$

For $\lambda \in \mathbb{R}$, we set

$$\Sigma_\lambda := \{x = (x^1, \dots, x^n) \in \mathbb{R}^n : x^1 < \lambda\}.$$

Given $x \in \mathbb{R}^n$, let $x^\lambda := (2\lambda - x^1, x^2, \dots, x^n)$ be the reflection of x about the plane $\partial\Sigma_\lambda = \{x^1 = \lambda\}$. We define

$$w_\lambda(x) := v(x^\lambda) - v(x), \quad \overline{w}_\lambda(x) := \frac{w_\lambda(x)}{g(x)}, \quad g(x) := \log(-x^1 + 3).$$

Notice that 0 is a possible singular point of v, and therefore $x_\lambda := (2\lambda, 0, \dots, 0)$ is a possible singular point of w_λ.

Lemma 5.9. *There exists $\lambda_0 \ll 0$ such that if*

$$\inf_{\widetilde{\Sigma}_\lambda} \overline{w}_\lambda < 0, \quad \widetilde{\Sigma}_\lambda := \Sigma_\lambda \setminus \{x_\lambda\},$$

for some $\lambda \le \lambda_0$, then the infimum is attained. Moreover, there exists $R_0 > 0$ (independent of λ) such that if x^0 is a minimum point of $\overline{w}_\lambda$ in $\widetilde{\Sigma}_\lambda$ with $\overline{w}_\lambda(x^0) < 0$ and $\lambda < 0$, then $|x^0| < R_0$.

Proof. Since $v > 0$ and $\Delta v \le 0$, we see that for $0 \ne x \in B_1$,

$$v(x) \ge \min_{\partial B_1} v =: \varepsilon_0 > 0.$$

Since $v(x) \to 0$ as $|x| \to \infty$, we can find $\lambda_0 \ll 0$ such that $v(x) \le \varepsilon_0$ in $B_1(x_\lambda)$ for $\lambda \le \lambda_0$. Therefore, for $\lambda \le \lambda_0$, if $\inf_{\widetilde{\Sigma}_\lambda} \overline{w} < 0$ then

$$\inf_{\widetilde{\Sigma}_\lambda} \overline{w}_\lambda = \inf_{\widetilde{\Sigma}_\lambda \setminus B_1(x_\lambda)} \overline{w}_\lambda,$$

and the above infimum is achieved.

In order to prove the second part of the lemma we first note that the function $v_\lambda(x) := v(x^\lambda)$ satisfies

$$-\Delta v_\lambda(x) \geq \frac{1}{|x|^\alpha} v_\lambda^q(x) \quad \text{on } \Sigma_\lambda,\ \lambda \leq 0.$$

It follows that

$$\Delta w_\lambda + c(x) w_\lambda \leq 0,$$

and therefore

$$\Delta \bar{w}_\lambda + \frac{2}{g} \nabla g \cdot \nabla \bar{w}_\lambda + \Big(c(x) + \frac{\Delta g}{g}\Big) \bar{w}_\lambda \leq 0 \quad \text{on } \Sigma_\lambda,\ \lambda \leq 0, \tag{5.3}$$

where $c(x) = q|x|^{-\alpha}\psi(x)^{q-1}$, and $\psi(x)$ is some number between $v(x)$ and $v(x^\lambda)$. It follows from the definition of v that, for some constants $c_1, c_2 > 0$,

$$\frac{c_1}{(1+|x|)^{n-2}} \leq v(x) \leq \frac{c_2}{(1+|x|)^{n-2}}.$$

Therefore, if $\bar{w}_\lambda(x) < 0$ and $\lambda \leq 0$, then necessarily

$$\frac{c_1}{(1+|x|)^{n-2}} \leq \frac{c_1}{(1+|x^\lambda|)^{n-2}} < v_\lambda(x) \leq \psi(x) \leq v(x) \leq \frac{c_2}{(1+|x|)^{n-2}}.$$

This implies that $c(x) \approx |x|^{-4}$ at infinity. Therefore, if $|x|$ is large then

$$c(x) + \frac{\Delta g(x)}{g(x)} < 0.$$

The second part of the lemma follows immediately from (5.3). ■

Proof of Theorem 5.8. It follows from Lemma 5.9 that for $\lambda \ll 0$ one has $\bar{w}_\lambda \geq 0$ in $\tilde{\Sigma}_\lambda$. Therefore, we can define

$$\bar{\lambda} := \sup\big\{\lambda < 0 \colon \bar{w}_\mu \geq 0 \text{ in } \tilde{\Sigma}_\mu\ \forall \mu \leq \lambda\big\}.$$

We claim that $\bar{\lambda} = 0$. Assuming by contradiction that $\bar{\lambda} < 0$, from the definition of $\bar{\lambda}$ we get that $\bar{w}_{\bar{\lambda}} \geq 0$ in $\tilde{\Sigma}_{\bar{\lambda}}$. Moreover, as $\bar{\lambda} < 0$ and $\alpha > 0$, it follows from (5.2) that $\bar{w}_{\bar{\lambda}} \not\equiv 0$ in $\tilde{\Sigma}_{\bar{\lambda}}$, and by the maximum principle, $\bar{w}_{\bar{\lambda}} > 0$ in $\tilde{\Sigma}_{\bar{\lambda}}$. Again, from the definition of $\bar{\lambda}$ we can find $\lambda_k \searrow \bar{\lambda}$ such that $\inf_{\tilde{\Sigma}_{\lambda_k}} \bar{w}_{\lambda_k} < 0$. Next we show that the infimum is achieved, that is

$$\inf_{\tilde{\Sigma}_{\lambda_k}} \bar{w}_{\lambda_k} = \bar{w}_{\lambda_k}(x_k)$$

for some $x_k \in \tilde{\Sigma}_{\lambda_k}$, and the sequence $(x_k)_k$ stays bounded away from $(x_{\lambda_k})_k$. Indeed, for some $\delta > 0$ and $\varepsilon > 0$ we have $\bar{w}_{\bar{\lambda}} \geq \varepsilon$ in $B_\delta(x_{\bar{\lambda}})$, which is a consequence of the fact that $\overline{w}_{\bar{\lambda}} > 0$ in $\widetilde{\Sigma}_{\bar{\lambda}}$ and $\Delta w_{\bar{\lambda}} \leq 0$. Therefore, from the convergence $\overline{w}_{\lambda_k} \to \overline{w}_{\bar{\lambda}}$ and Lemma 5.9, the minimum point x_k exists, (x_k) is bounded, and x_k stays bounded away from x_{λ_k}. This implies that, up to a subsequence, $x_k \to \bar{x} \neq x_{\bar{\lambda}}$, $\overline{w}_{\bar{\lambda}}(\bar{x}) = 0$, and $\nabla \overline{w}_{\bar{\lambda}}(\bar{x}) = 0$. As $\overline{w}_{\bar{\lambda}} > 0$ in $\widetilde{\Sigma}_{\bar{\lambda}}$, we necessarily have that $\bar{x} \in \partial \Sigma_{\bar{\lambda}}$, and one obtains a contradiction by Hopf's lemma. Thus, $\bar{\lambda} = 0$.

From the above arguments we see that

$$v(-x^1, x^2, \dots, x^n) \geq v(x^1, x^2, \dots, x^n) \quad \text{for } x^1 < 0. \tag{5.4}$$

Repeating the previous procedure in the opposite direction, namely moving the parallel planes in the negative x^1-direction from $+\infty$ towards the origin, one would obtain a reverse inequality in (5.4). In conclusion, v is symmetric about the plane $\{x^1 = 0\}$. Since the x^1-direction is arbitrary, we conclude that v is radially symmetric, and consequently, u is radially symmetric. However, as we can choose the origin arbitrarily, we see that $u \equiv 0$. ∎

Chapter 6

Blow-up analysis

Our main goal is to derive an existence theory for the conformal prescribed curvature equation on a closed manifold (M, g) of positive Yamabe class, namely

$$L_g u = -c_n \Delta_g u + R_g u = K u^{\frac{n+2}{n-2}}. \tag{E$_K$}$$

We will always assume that K is a positive smooth Morse function, i.e.

$$\nabla K(x) = 0 \Rightarrow \nabla^2 K(x) \text{ is non-degenerate},$$

where we denote by $\nabla^2 K(x)$ the Hessian of K at x; see Chapter 2. This implies in particular that critical points of K are finitely many.

We will also assume the *non-degeneracy condition* (ND), which we recall:

$$\nabla K(x) = 0 \Rightarrow \Delta K(x) \neq 0. \tag{ND}$$

It is worth mentioning that the two above conditions, even together, are generic: functions satisfying them are dense in C^2, and also form an open set; see e.g. [177].

We saw that solutions of (E$_K$) can be found as critical points of the Euler–Lagrange functional J_K in (4.3), but that, in general, minimizing or Palais–Smale sequences may diverge; see Proposition 4.3 and Theorem 4.6. It will be then convenient to introduce a *sub-critical approximation* of (E$_K$), in the spirit of a well-known method developed by Sacks–Uhlenbeck [192] for harmonic maps, also attempted by Yamabe [218].

For $\tau > 0$ small, consider the problem

$$L_g u = K u^{\frac{n+2}{n-2}-\tau} \quad \text{on } (M, g). \tag{E$_{K,\tau}$}$$

An advantage of (E$_{K,\tau}$) is that it is always solvable. To see this, let us consider the corresponding Euler–Lagrange functional

$$J_{K,\tau}(u) := \frac{\int_M [c_n |\nabla u|^2 + R_g u^2] \, dV_g}{\left(\int_M K |u|^{2^*-\tau} \, dV_g\right)^{\frac{2}{2^*-\tau}}}. \tag{6.1}$$

The assumption $\lambda_1(L_g) > 0$ implies that the numerator is equivalent to $\|u\|^2_{H^1(M)}$. As $H^1(M)$ embeds compactly in $L^{2^*-\tau}(M)$, minimizing sequences for $J_{K,\tau}$ (normalized, say, in $L^{2^*-\tau}(M)$) sub-converge in $H^1(M)$ to a global minimizer $\hat{u}$, solving

$$L_g \hat{u} = \mu K \hat{u}^{2^*-\tau}.$$

As before, one can always multiply $\hat{u}$ by a proper constant to get $\mu = 1$, implying existence for $(E_{K,\tau})$.

A disadvantage of $(E_{K,\tau})$ is that its solutions may blow up as $\tau \to 0$. For example, take $K(y) = 2 + y^{n+1}$ on $\mathbb{S}^n$ with $y = (y^1, \ldots, y^{n+1}) \in \mathbb{S}^n \subseteq \mathbb{R}^{n+1}$. Such a function possesses a non-degenerate maximum, a non-degenerate minimum, it is Morse, and satisfies (ND). However, K is *forbidden* by the Kazdan–Warner identity (see Section A.2), i.e. problem (E_K) is not solvable for this choice of K. Recall that minimizing sequences for J_K concentrate at a single point where $K_{\max}$ is realized; see the comment after Proposition 4.3. In fact, as we will see in the next chapter, there are solutions u_τ that diverge at such points for $\tau \to 0$.

Solutions of $(E_{K,\tau})$ sometimes behave qualitatively like a P-S sequence for J_K. However, they are exact solutions of an equation and carry therefore more information. For example, in three dimensions, if (u_τ) is a family of blowing-up (positive) solutions to $(E_{K,\tau})$ (this means $\limsup_{\tau\to 0} \max_M u_\tau = \infty$), then $u_\tau \rightharpoonup 0$ and u_τ blows up at only one point $\bar{x} \in \mathbb{S}^3$ (with a bubble profile). Moreover, $\nabla K(\bar{x}) = 0$ and $\Delta K(\bar{x}) < 0$. On the other hand, it is easy to construct Palais–Smale sequences for J_K as in Theorem 4.6 with non-zero weak limit and multiple bubbles that can accumulate at any given point of M.

6.1 Isolated and isolated-simple blow-ups

As we will see shortly, by exploiting the classification result Theorem 5.1, a blowing-up sequence of solutions will always develop a *bubble profile*, characterizing blowing-up solutions in some shrinking metric balls. The next goal would then be to try extending the estimates to some larger sets, so we will give definitions useful for this purpose.

We will work for simplicity only in $\mathbb{R}^n$: the case of general manifolds requires similar ideas but more technical details. We will explicitly comment when this will make a concrete difference. In the case of the round sphere $\mathbb{S}^n$, equation $(E_{K,\tau})$ for $\tau = \tau_i \to 0$ can be written in convenient coordinates using the stereographic projection and conformal covariance formulas as in the proof of Theorem 3.4, and becomes

$$-\Delta u_i = U(x)^{\tau_i} K(x) u_i^{\frac{n+2}{n-2}-\tau_i}, \tag{6.2}$$

where we are composing K with the inverse stereographic projection, and the extra factor $U(x)^{\tau_i}$ (see Chapter 3) arises from the sub-criticality of the exponent. For this reason, when we let $\tau \to 0$ along a sequence $(\tau_i)_i$, it will be convenient to consider K as depending on i as well, and converging to the original curvature function to be prescribed.

The next proposition yields the claimed limiting bubble profile, after proper rescaling, of solutions defined on the open unit ball B_1 in $\mathbb{R}^n$.

Proposition 6.1. *Let $\tau_i \to 0$, $K_i \to K$ in $C^2(\bar{B}_1)$, and let x_i be a maximum point of u_i in B_1, solving*

$$-\Delta u_i = \frac{1}{c_n} K_i u_i^{p_i} \text{ on } B_1, \quad p_i := \frac{n+2}{n-2} - \tau_i, \tag{E_{K_i,τ_i}}$$

with $x_i \to \bar{x} \in B_1$. Define

$$v_i(x) := M_i^{-1} u_i\big(M_i^{-\frac{p_i-1}{2}} x + x_i\big), \quad M_i := u_i(x_i). \tag{6.3}$$

Then

$$v_i \to (1 + \bar{K}|x|^2)^{-\frac{n-2}{2}} \text{ in } C^2_{\mathrm{loc}}(\mathbb{R}^n), \quad \bar{K} := (c_n n(n-2))^{-1} K(\bar{x}).$$

Proof. The function v_i (defined on domains invading $\mathbb{R}^n$) satisfies

$$-\Delta v_i = \frac{1}{c_n} K_i\big(M_i^{-\frac{p_i-1}{2}} x + x_i\big) v_i^{p_i}, \quad 0 < v_i \le v_i(0) = 1, \ \nabla v_i(0) = 0.$$

Using elliptic regularity theory it is easy to prove that, up to a subsequence, $v_i \to \bar{v}$ in $C^2_{\mathrm{loc}}(\mathbb{R}^n)$, where $\bar{v}$ solves

$$-\Delta \bar{v} = \frac{1}{c_n} K(\bar{x}) \bar{v}^{\frac{n+2}{n-2}} \text{ on } \mathbb{R}^n, \quad \bar{v} \le \bar{v}(0) = 1, \ \nabla \bar{v}(0) = 0.$$

By the classification result in Theorem 5.1 one gets

$$\bar{v} = (1 + \bar{K}|x|^2)^{-\frac{n-2}{2}},$$

which concludes the proof. ∎

As a consequence of the above theorem we see that for some $R_i \to \infty$ (slowly enough) one has

$$\|v_i - (1 + \bar{K}|\cdot|^2)^{-\frac{n-2}{2}}\|_{C^2(B_{R_i})} \to 0 \quad \text{as } i \to \infty.$$

Therefore, scaling back in (6.3), we have control on u_i in the balls $B_{R_i M_i^{-(p_i-1)/2}}(x_i)$, whose radii are tending to zero.

Definition 6.2. Let (u_i) solve $(\mathrm{E}_{K_i,\tau_i})$ in B_1 for some $\tau_i \to 0$ and let $K_i \to K > 0$ in $C^2(\bar{B}_1)$. A point $\bar{x} \in B_1$ is called a *blow-up point* if there exists $x_i \to \bar{x}$ such that $u_i(x_i) \to \infty$. The point $\bar{x}$ is called an *isolated blow-up* point if there exist $\bar{c} > 0$, $\bar{r} > 0$, $\bar{r} < \operatorname{dist}(\bar{x}, \partial B_1)$, and there exists a sequence of local maxima $(x_i)_i$ for $(u_i)_i$ such that for i large,

$$x_i \to \bar{x}, \quad u_i(x_i) \to \infty, \quad \text{and} \quad u_i(x) \le \bar{c}|x - x_i|^{-\frac{2}{p_i-1}} \quad \text{for } x \in B_{\bar{r}}(x_i).$$

Remark 6.3. A heuristic interpretation of the above definition, for $\tau = 0$, is as follows. Let U be a bubble profile and recall its scaled version $U_\lambda(x) = \lambda^{\frac{n-2}{2}} U(\lambda x)$. The function $|x|^{-\frac{n-2}{2}}$ is the envelope of the U_λ, i.e. $\sup_{\lambda>0} U_\lambda(x)$; see Figure 6.1. For isolated blow-ups the formation of multiple bubbles is allowed, but they should all have nearby centers.

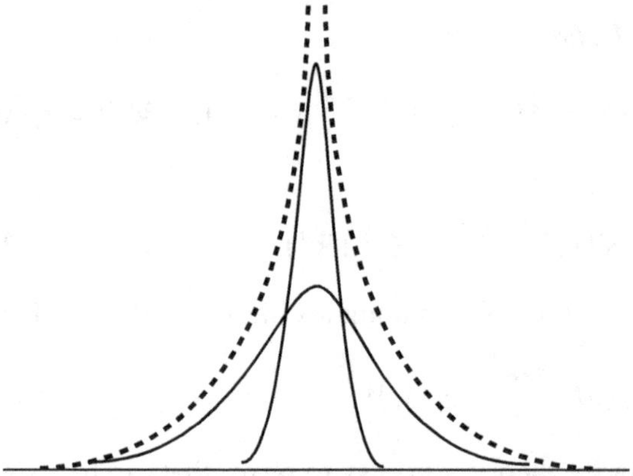

Figure 6.1. Envelope of bubbles with the same center and different scales

Solutions developing isolated blow-ups satisfy the following useful property.

Lemma 6.4 (Harnack inequality). *Let $\bar{x}$ be an isolated blow-up point. Then there exists $C = C(n, \bar{c}, \|K_i\|_{L^\infty})$ such that for every $r \in (0, \frac{\bar{r}}{3})$ one has*

$$\sup_{B_{2r}(x_i) \setminus B_{\frac{r}{2}}(x_i)} u_i \leq C \inf_{B_{3r}(x_i) \setminus B_{\frac{r}{3}}(x_i)} u_i .$$

Notice that in the previous statement r can be arbitrarily small, and that the constant C does not depend on r.

Proof of Lemma 6.4. Defining

$$v_i(x) := r^{\frac{2}{p_i - 1}} u_i(x_i + rx),$$

we see that v_i satisfies

$$-\Delta v_i = \frac{1}{c_n} K_i(x_i + rx) v_i^{p_i}(x) \text{ on } B_3, \quad 0 < v_i(x) < \bar{C}|x|^{-\frac{2}{p-1}} \text{ on } B_3.$$

Then one can apply the standard Harnack inequality (see e.g. [113, Theorem 8.8]) for v_i in $B_{\frac{9}{4}} \setminus B_{\frac{1}{4}}$, to deduce the conclusion. ∎

A consequence of the previous lemma is that if $\bar{x}$ is an isolated blow-up point for $(u_i)_i$, then u_i is controlled by its radial average (by the Harnack inequality). Define

$$\bar{u}_i(r) := \frac{1}{|\partial B_r|} \int_{\partial B_r(x_i)} u_i \, d\sigma, \quad \bar{w}_i(r) := r^{\frac{2}{p-1}} \bar{u}_i(r), \quad r > 0. \tag{6.4}$$

Definition 6.5. An isolated blow-up point $\bar{x}$ is called *isolated simple* if there exists $\rho > 0$ such that $\bar{w}_i(\cdot)$ has precisely one critical point in $(0, \rho)$ for i large.

Remark 6.6. Similarly to the previous remark, we can heuristically interpret the last definition as follows, for $\tau = 0$: Recall the effect of conformal transformations on conformal factors as in the proof of Theorem 3.4. Then consider the cylindrical metric obtained by conformal deformation of the Euclidean one on $\mathbb{R}^n \setminus \{x_i\}$, namely

$$|x - x_i|^{-2} \, dx^2 = d(\log|x - x_i|)^2 + d\theta^2,$$

where θ is the polar variable around x_i. Then define $\tilde{u}_i(x) := |x - x_i|^{\frac{n-2}{2}} u_i(x)$. While the condition of isolated blow-up gives the boundedness of $\tilde{u}_i$ on $B_{\bar{r}}(x_i)$, the extra property of an (isolated) simple blow-up means that the radial average of $\tilde{u}_i$ has a single critical point for radii in $(0, \rho)$.

If u_i develops multiple bubbles at the same point, still under an isolated blow-up condition, the function $\tilde{u}_i$, represented in Figure 6.2 by logarithmic coordinates, would have multiple oscillations, so in this case the blow-up could not be isolated simple.

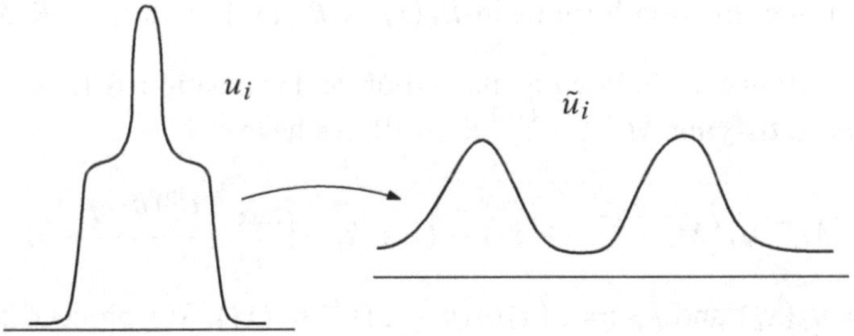

Figure 6.2. A non-isolated-simple blow-up and its cylindrical representative

We state next a consequence of the isolated simple blow-up condition, somehow related to the discussion in Remark 3.15.

Proposition 6.7. *Suppose u_i satisfies $(\mathrm{E}_{K_i,\tau_i})$ (with K_i as before) in B_2, and that $x_i \to 0$ is an isolated simple blow-up with $\bar{r} = 2$ (according to Definition 6.2). Then*

there exists $b\colon B_1 \to \mathbb{R}$ *harmonic such that, up to a subsequence,*

$$u_i(x_i)u_i(x) \to h(x) := a|x|^{2-n} + b(x) \ \text{in}\ C^2_{\mathrm{loc}}(B_1 \setminus \{0\}),$$
$$a := \Big(\frac{n(n-2)c_n}{K(0)}\Big)^{\frac{n-2}{2}}.$$

Furthermore,

$$u_i(x_i)^{\tau_i} = 1 + o_i(1) \ \text{as}\ i \to \infty \quad \text{and} \quad u_i(x) \le C u_i(x_i)^{-1}|x - x_i|^{2-n} \ \text{on}\ B_1.$$

Corollary 6.8. *Suppose* 0 *is an isolated simple blow-up for* $(u_i)_i$, *and let* $C_{K(0)}$ *be such that*

$$-\Delta(C_{K(0)}U) = K(0)(C_{K(0)}U)^{\frac{n+2}{n-2}}.$$

Then, for $r > 0$ *small and fixed, one has in* $B_r(x_i)$,

$$u_i(x) = C_{K(0)}U_{\lambda_i}(x - x_i)(1 + o_i(1) + o_r(1)),$$
$$\nabla u_i(x) = C_{K(0)}\nabla U_{\lambda_i}(x - x_i)(1 + o_i(1) + o_r(1)),$$

with

$$U_\lambda(x) := \lambda^{\frac{n-2}{2}}U(\lambda x), \quad U(x) = (1 + |x|^2)^{-\frac{n-2}{2}},$$
$$C_{K(0)}\lambda_i^{\frac{n-2}{2}} = u_i(x_i).$$

Using the upper bound $u_i(x_i)u_i(x) \le C|x - x_i|^{2-n}$, the corollary can be obtained from Green's representation formula in $B_r(x_i) \setminus B_{r_i}(x_i)$, with $r_i = R_i M_i^{-(p_i-1)/2}$.

Proof of Proposition 6.7. Following the proof of Proposition 6.1, we see that for some $R_i \to \infty$ satisfying $M_i^{-(p_i-1)/2}R_i \to 0$, we have

$$M_i^{-1}u_i\big(M_i^{-\frac{p_i-1}{2}} \cdot + x_i\big) - (1 + k_i|\cdot|^{\frac{2-n}{2}}) \xrightarrow{C^2(B_{2R_i})} 0, \tag{6.5}$$

where $M_i := u_i(x_i)$ and $k_i := c(n)(n(n-2))^{-1}K_i(x_i)$. We proceed then in three steps.

Step 1. We have

$$u_i(x) \le C u_i(x_i)^{-1+o_i(1)}|x - x_i|^{2-n+o_i(1)} \quad \text{for } R_i u_i(x_i)^{-\frac{p_i-1}{2}} \le |x - x_i| \le 1.$$

In order to prove this fact, we notice that

$$u_i(x) \le C u_i(x_i)R_i^{2-n} \quad \text{for } |x - x_i| = r_i := R_i u_i(x_i)^{-\frac{p_i-1}{2}}. \tag{6.6}$$

Let $\bar{u}_i(\cdot)$ be the radial average of u_i as defined in (6.4). It follows from the assumption of isolated simple blow-up and (6.5) that

$$\bar{w}_i(r) := r^{\frac{2}{p_i-1}}\bar{u}_i(r) \quad \text{is strictly monotone decreasing on } (r_i, \rho), \tag{6.7}$$

where $\rho > 0$ is as in Definition 6.5. Together with Harnack's inequality and (6.6)–(6.7) one obtains

$$|x - x_i|^{\frac{2}{p_i-1}} u_i(x) \leq C R_i^{\frac{2-n}{2}+o(1)} \quad \text{for } r_i < |x - x_i| < \rho,$$

which gives

$$u_i(x)^{p_i-1} \leq C R_i^{-2+o(1)} |x - x_i|^{-2} \quad \text{for } r_i < |x - x_i| < \rho.$$

Note that the linear operator

$$L_i\varphi := \Delta\varphi + \frac{1}{c_n} K_i u_i^{p-1}\varphi$$

satisfies the maximum principle, since $L_i u_i = 0$ with $u_i > 0$. It is possible to find a sequence $\delta_i = O(R_i^{-2+o(1)}) \to 0$ such that

$$L_i(|x - x_i|^{-\delta_i}) \leq 0, \quad L_i(|x - x_i|^{2-n+\delta_i}) \leq 0 \quad \text{for } r_i \leq |x - x_i| \leq \rho.$$

Then, setting $\lambda_i := \frac{1}{2}(2 - n - \delta_i)(p_i - 1) - 1$, $\Lambda_i := \max_{\partial B_\rho} u_i$, and (the constant $A \gg 1$ will be chosen later)

$$\varphi_i(x) := \Lambda_i \rho^i |x - x_i|^{-\delta_i} + A u_i(x_i)^{-\lambda_i} |x - x_i|^{2-n+\delta_i},$$

we see that φ_i is a super-solution of L_i in $r_i \leq |x - x_i| \leq \rho$. In order to use the maximum principle, we observe that

$$\varphi_i \geq \Lambda_i \geq u_i \quad \text{if } |x - x_i| = \rho,$$

and

$$\varphi_i \geq A u_i(x_i) R_i^{2-n+\delta_i} \geq A u_i(x_i) R_i^{2-n} \geq u_i \quad \text{if } |x - x_i| = r_i,$$

for $A \gg 1$, thanks to (6.6). Therefore, it follows that

$$u_i \leq \varphi_i \quad \text{for } r_i \leq |x - x_i| \leq \rho. \tag{6.8}$$

This, (6.7), and Harnack's inequality imply

$$\rho^{\frac{p_i-1}{2}} \Lambda_i \leq C\theta^{\frac{p_i-1}{2}} (\Lambda_i \rho_i^{\delta} \theta^{-\delta_i} + A u_i(x_i)^{-\lambda_i} \theta^{2-n+\delta_i}),$$

for any $r_i < \theta < \rho$. Choosing $\theta > 0$ sufficiently small verifying

$$C\theta^{\frac{p_i-1}{2}} \rho^{\delta_i} \theta^{-\delta_i} \leq \frac{1}{2}\rho^{\frac{p_i-1}{2}},$$

one gets that $\Lambda_i \leq C u_i(x_i)^{-\lambda_i}$. Step 1 follows from (6.8) and Harnack's inequality.

Step 2. We have

$$\tau_i = O(u_i(x_i)^{-\frac{2}{n-2}+o_i(1)}), \quad \text{and in particular } u_i(x_i)^{\tau_i} = 1 + o_i(1).$$

Let $\mathcal{B}$ be as in (A.1): by Corollary A.2 one can show that (for $r < 1$)

$$\int_{\partial B_r(x_i)} \mathcal{B}\, d\sigma = c_{n,p} \int_{B_r(x_i)} \langle x, \nabla K_i \rangle u_i^{p_i+1}\, dx + \bar{c}_n \tau_i \int_{B_r(x_i)} K_i u_i^{p_i+1}\, dx \\ - r\bar{c}_{n,p_i} \int_{\partial B_r(x_i)} K_i u_i^{p_i+1}\, d\sigma,$$

which implies

$$\int_{\partial B_r(x_i)} \mathcal{B}\, d\sigma = O(u_i(x_i)^{-\frac{2}{n-2}+o_i(1)}) + \bar{c}_n \tau_i + \text{l.o.t.}$$

Step 2 follows from (6.5), Step 1, and Harnack's inequality.

Step 3. Consider $e \in \mathbb{R}^n$ with $|e| = 1$, and set $v_i = u_i(x_i + e)^{-1} u_i(x)$. Then

$$v_i \to v \text{ in } C^{2,\alpha}_{\text{loc}}(B_2 \setminus \{0\}), \quad v(x) := a_1 |x|^{2-n} + b_1(x).$$

It follows that $u_i(x_i + e) \to 0$, and the function v_i satisfies

$$-\Delta v_i = \frac{1}{c_n} K_i(x) u_i(x_i + e)^{p_i - 1} v_i^{p_i} \to 0 \quad \text{in } C^{\alpha}_{\text{loc}}(B_{\frac{3}{2}} \setminus \{0\}),$$

thanks to Step 1. Therefore, by standard elliptic theory, up to a subsequence $v_i \to v$ in $C^{2,\alpha}_{\text{loc}}(B_2 \setminus \{0\})$, where v satisfies

$$\Delta v = 0 \quad \text{on } B_2 \setminus \{0\}.$$

Moreover, by Harnack's inequality,

$$u_i(x_i + e)^{-1} r^{\frac{2}{p_i - 1}} \bar{u}_i(r) \to r^{\frac{n-2}{2}} \bar{v}(r) \quad \text{for every } r \in (0, 2),$$

where we have set $\bar{v}(r) = \frac{1}{|\partial B_r|} \int_{\partial B_r} v\, d\sigma$. From the definition of isolated simple blow-up, we deduce that the function

$$r \mapsto r^{\frac{n-2}{2}} \bar{v}(r)$$

is monotone decreasing on $(0, \rho)$ (we can assume for simplicity that $\rho \in (0, \frac{1}{2})$). This shows that $\bar{v}$ has a non-removable singularity at the origin, and consequently

$$v(x) = a_1 |x|^{2-n} + b_1(x),$$

for some constant $a_1 > 0$ and a harmonic function b_1 defined on B_2.

We are now in position to complete the proof of Proposition 6.7. We start by showing

$$u_i(x_i + e) \le C u_i(x_i)^{-1}. \tag{6.9}$$

In order to prove (6.9), we assume by contradiction that for some subsequence,

$$u_i(x_i + e) u_i(x_i) \to \infty.$$

Then, multiplying the equation $-\Delta u_i = \frac{1}{c_n} K_i u_i^{p_i}$ by $u_i(x_i + e)$ and integrating the resultant on B_1, we obtain

$$-\int_{\partial B_1} \partial_\nu v_i \, d\sigma = \frac{1}{c_n} u_i(x_i + e)^{-1} \int_{B_1} K_i u_i^{p_i} \, dx. \tag{6.10}$$

The harmonicity of b_1 implies

$$\lim_{i \to \infty} \int_{\partial B_1} \partial_\nu v_i \, d\sigma = a_1 \int_{\partial B_1} \partial_\nu |x|^{2-n} \, d\sigma + \int_{\partial B_1} \partial_\nu b_1 \, d\sigma < 0.$$

Concerning the integral on the right-hand side of (6.10), by Step 2 and (6.5),

$$u_i(x_i + e)^{-1} \int_{|x - x_i| \le r_i} K_i u_i^{p_i} \, dx \le C u_i(x_i + e)^{-1} u_i(x_i)^{-1} \to 0,$$

and by Steps 1 and 2,

$$\begin{aligned} \int_{r_i \le |x - x_i| \le 1} K_i u_i^{p_i} \, dx &\le C \int_{r_i \le |x - x_i| \le 1} (u_i(x_i)^{-\lambda_i} |x - x_i|^{2-n+\delta_i})^{p_i} \, dx \\ &\le C R_i^{n - p_i(n-2-\delta_i)} u_i(x_i)^{-1+\tau_i \frac{n-2}{2}} \\ &= o(1) u_i(x_i)^{-1}. \end{aligned}$$

Thus, the right-hand side of (6.10) tends to 0, a contradiction.

As consequence of (6.9) and of Harnack's inequality, we obtain

$$u_i(x) \le C u_i(x_i)^{-1} |x - x_i|^{2-n} \quad \text{for } \rho \le |x - x_i| \le 1.$$

In order to complete the proof of Proposition 6.7 we only need to establish the above inequality in the region $r_i \le |x - x_i| \le \rho$, and compute the value of the constant a.

We assume by contradiction that there exists $\tilde{x}_i$ satisfying $r_i \le |x_i - \tilde{x}_i| \le \rho$ such that

$$u_i(\tilde{x}_i) u_i(x_i) |x - x_i|^{n-2} \to \infty.$$

Setting

$$\tilde{u}_i(x) := (\tilde{r}_i)^{\frac{2}{p_i - 1}} u_i(x_i + r_i x), \quad \tilde{r}_i := |\tilde{x}_i - x_i|,$$

we see that $\tilde{u}_i$ satisfies

$$-\Delta \tilde{u}_i = K_i(x_i + \tilde{r}_i x)\tilde{u}_i^{p_i} \quad \text{on } B_2,$$

and all the hypotheses of Proposition 6.7, with 0 being an isolated simple blow-up point. It then follows from (6.9) that

$$\tilde{u}_i(0)\tilde{u}_i\Big(\frac{\tilde{x}_i - x_i}{\tilde{r}_i}\Big) \le C,$$

and together with Step 2,

$$u_i(\tilde{x}_i)u_i(x_i)|\tilde{x}_i - x_i| \le C.$$

This is a contradiction, and we conclude the proof of the inequality in Proposition 6.7.

To compute the value of the constant a, one can multiply the equation for u_i by $u_i(x_i)$ and integrate over B_1. The fact that the function b_1 is harmonic, estimate (6.5), and Step 1 lead to the desired conclusion. ∎

6.2 Blow-up analysis in dimensions 3 and 4

As for the previous section, we will only work on $\mathbb{R}^n$ (or on $\mathbb{S}^n$, by (6.2)), commenting on the general case at specific points. In low dimensions, we have the following result.

Theorem 6.9 ([57, 151, 157]). *Let (M^n, g) be compact, $\lambda_1(L_g) > 0$, $K_i \to K > 0$ in $C^2(M)$, $\tau_i \to 0$, $\tau_i \ge 0$. Let $u_i > 0$ be solutions of* $(\mathrm{E}_{K,\tau})$ *with $K = K_i, \tau = \tau_i$, and satisfying* $\max_M u_i \to \infty$. *If $n = 3, 4$ then u_i has only (finitely many) isolated simple blow-ups at critical points of K with $\Delta K < 0$, $\|u_i\|_{H^1(M)} \le C$, and $u_i \rightharpoonup 0$. If $n = 3$, there is at most one blow-up point, and if (M^3, g) is not conformally equivalent to $(\mathbb{S}^3, g_{\mathbb{S}^3})$ then there is no blow-up.*

The strategy will consist in proving that for $n = 3, 4$, isolated blow-ups are indeed isolated simple, and then that all blow-ups are isolated. The results in Theorem 6.9 have some previous counterparts in [20, 23, 24] via the theory of *critical points at infinity*, briefly discussed in Chapter 10. We choose here to follow the approach via blow-up theory, which is static in nature, compared to the dynamical flavor of the other one.

Proposition 6.10 (See [194] for $K \equiv 1$). *Suppose that $n = 3, 4$ and that $x_i \to 0$ is an isolated blow-up with $\bar{r} = 2$. Then 0 is isolated simple.*

Proof. Recall that, by Proposition 6.1, $\overline{w}_i(r) = r_i^{2/(p_i-1)}\overline{u}_i(r)$ has a critical point at scale $r_i = M_i^{-(p_i-1)/2} \to 0$. Suppose by contradiction that there exists a second *next* critical point for $\overline{w}_i$ at scale $r = s_i \to 0$, $s_i \gg r_i$. Consider

$$\tilde{u}_i(x) = s_i^{\frac{2}{p_i-1}} u_i(s_i x).$$

Then $\tilde{u}_i$ has maximum points $\tilde{x}_i$ (without loss of generality, we can assume $\tilde{x}_i \to 0$) and 0 is an isolated blow-up with $\bar{r}_i \to \infty$ (i.e. $\tilde{u}_i(x) \leq \bar{C}|x - \tilde{x}_i|^{2/(p_i-1)}$ in $B_{\bar{r}_i}(0)$). By definition of s_i, the origin is also an isolated simple blow-up point, with $\widetilde{w}_i'(1) = 0$, $\widetilde{w}_i(r) = r^{2/(p_i-1)}\bar{\tilde{u}}_i(r)$.

Since $\bar{r}_i \to \infty$, we have radial Harnack inequalities on expanding annuli, by Lemma 6.4. By Proposition 6.7, for some harmonic function $\tilde{b}$ there holds

$$\tilde{u}_i(\tilde{x}_i)\tilde{u}_i(x) \to \tilde{a}|x|^{2-n} + \tilde{b}(x), \quad \tilde{a} > 0. \tag{6.11}$$

Moreover, $\tilde{u}_i(\tilde{x}_i)\tilde{u}_i(x)$ satisfies

$$-c_n\Delta(\tilde{u}_i(\tilde{x}_i)\tilde{u}_i) = K(s_i x)(\tilde{u}_i(\tilde{x}_i)\tilde{u}_i)^{p_i}\frac{1}{\tilde{u}_i(\tilde{x}_i)^{p_i}}.$$

As the right-hand side tends to 0 as $i \to +\infty$, the sequence $\tilde{u}_i(\tilde{x}_i)\tilde{u}_i$ converges by the Harnack inequality in $C^2_{\mathrm{loc}}(\mathbb{R}^n \setminus \{0\})$. So, the convergence in (6.11) holds in $C^2_{\mathrm{loc}}(\mathbb{R}^n \setminus \{0\})$. It follows that $\tilde{b} \geq -C$, and hence by Liouville's theorem $\tilde{b}$ is a constant function. The fact that $\widetilde{w}_i'(1) = 0$ implies

$$\frac{d}{dr}(r^{\frac{n-2}{2}}(\tilde{a}r^{2-n} + \tilde{b}))|_{r=1} = 0 \;\Rightarrow\; \tilde{a} = \tilde{b} > 0.$$

We now use the Pohozaev identity of Corollary A.2 together with Corollary 6.8 in $B_r(\tilde{x}_i)$ to obtain

$$\begin{aligned} c_{n,p}\int_{B_r(x_i)}\langle x, \nabla K_i(s_i x)\rangle \tilde{u}_i^{p_i+1}\,dx + \bar{c}_n\tau\int_{B_r(x_i)} K_i\tilde{u}_i^{p_i+1}\,dx \\ - r\bar{c}_{n,p_i}\int_{\partial B_r(x_i)} K_i\tilde{u}_i^{p_1+1}\,d\sigma = \int_{\partial B_r(x_i)} \mathcal{B}\,d\sigma. \end{aligned}$$

By a Taylor expansion of $K_i(s_i\cdot)$ (similar to (4.9)), and using the fact that the first-order term vanishes by oddness, we get

$$\int_{B_r(x_i)}\langle x, \nabla K_i(s_i x)\rangle\tilde{u}_i^{p_i+1}\,dx \approx s_i\frac{\Delta K(0)}{\tilde{\lambda}_i^2}, \qquad \int_{B_r(x_i)} K_i\tilde{u}_i^{p_i+1}\,dx \approx 1,$$

$$\int_{\partial B_r(x_i)} K_i\tilde{u}_i^{p_1+1}\,d\sigma = O(\tilde{u}_i(\tilde{x}_i)^{-2^*}), \qquad \int_{\partial B_r(x_i)} \mathcal{B}\,d\sigma \approx -\tilde{b}\tilde{u}_i(\tilde{x}_i)^{-2}.$$

Recalling that $\tilde{u}_i(\tilde{x}_i) = \tilde{\lambda}_i^{\frac{n-2}{2}}$ one finds an identity of the following type, up to positive coefficients in each summand

$$s_i \Delta K(0)\tilde{\lambda}_i^{-2} + \tau + \tilde{b}\lambda_i^{-(n-2)} = O(\tilde{\lambda}_i^{-n}),$$

a contradiction of the fact that $\tilde{b} = \tilde{a} > 0$ in case $n = 3$ or $n = 4$. ∎

Remark 6.11. We have the following extra properties:

(i) Once we know that the blow-up is isolated simple, we can repeat the above argument (from (6.11) onward) directly for u_i in dimension $n = 3$ or 4. If $x_i^k \to \bar{x}^k$ are isolated simple blow-up points, then $u_i(x_i^k)u_i(x)$ converges in $C^2_{\mathrm{loc}}(B_r(\bar{x}^k) \setminus \{\bar{x}^k\})$ to a positive singular solution h of $L_g h = 0$ on $M \setminus \bigcup_{j=1}^L \bar{x}^j$. By Harnack's inequality and the first statement in Proposition 6.7, $u_i(x_i^k) \approx u_i(x_i^j)$ for $j \neq k$. On M, $h = \sum_{j=1}^L \alpha_j G_{\bar{x}^j}$, where $G_{\bar{x}^j}$ is the Green's function of L_g with pole at $\bar{x}^j$, and $\alpha_j > 0$. The term $b(0)$ (as in Proposition 6.7) at a given blow-up point $\bar{x} = \bar{x}^1$ is replaced by $\alpha_1 A(\bar{x}) + \sum_{j=2}^L \alpha_j G_{\bar{x}^j}(\bar{x})$, where A is the *Robin function* as in Remark 3.15. Then the Pohozaev identity yields, up to some positive coefficient in each summand,

$$\Delta K(\bar{x})\lambda_i^{-2} + \tau + \left(\alpha_1 A(\bar{x}) + \sum_{j=2}^{L} \alpha_j G_{\bar{x}^j}(\bar{x})\right)\lambda_i^{-(n-2)} = O(\lambda_i^{-n}). \quad (6.12)$$

(ii) In the case of isolated simple blow-ups, it is possible to show also that $\nabla K_i(x_i) = o_i(1)$, and hence $\nabla K(\bar{x}) = 0$. To see this, multiply equation $(\mathrm{E}_{K_i,\tau_i})$ by $\eta \frac{\partial u_i}{\partial x^j}$, where η is a cutoff function such that $\eta = 1$ in $B_{\frac{1}{4}}(\bar{x})$, $\eta = 0$ outside $B_{\frac{1}{2}}(\bar{x})$. Without loss of generality we assume that $\bar{x} = 0$. Integrating by parts one finds that

$$\begin{aligned}\frac{c_n^{-1}}{p_i+1}\int_{B_1} \frac{\partial K_i}{\partial x^j} u_i^{p_i+1}\eta\, dx &= \frac{1}{2}\int_{B_{\frac{1}{2}}\setminus B_{\frac{1}{4}}} |\nabla u_i|^2 \frac{\partial \eta}{\partial x^j}\, dx \\ &\quad - \int_{B_{\frac{1}{2}}\setminus B_{\frac{1}{4}}} \frac{\partial u_i}{\partial x^j}\langle \nabla u_i, \nabla \eta\rangle\, dx \\ &\quad - \frac{c_n^{-1}}{p_i+1}\int_{B_{\frac{1}{2}}\setminus B_{\frac{1}{4}}} K_i u_i^{p_i+1}\frac{\partial \eta}{\partial x^j}\, dx.\end{aligned}$$

The left-hand side converges to $\hat{c}_n \nabla K(0)$ for some $\hat{c}_n > 0$, while the right-hand side converges to zero, giving the desired conclusion.

We state next a useful result due to Schoen, regarding solutions of $(\mathrm{E}_{K,\tau})$ for τ small on a closed manifold M. As we will see, M can be divided into small balls where solutions have a bubble profile, while in their complement an estimate in the spirit of Definition 6.2 is satisfied. A priori, the number of the above balls depends on the solution itself, but we will see that in low dimensions this number is actually uniformly bounded.

Proposition 6.12 ([194]). *Suppose $\tau \geq 0$ and that $K \in C^1(M,g)$ satisfies $K \in [\frac{1}{A}, A]$, $\|\nabla K\|_{L^\infty} \leq A$. Then, for every $\varepsilon > 0$ and $R \geq 1$, there exist large constants C_0, $C_1 = C_i(\varepsilon, R, M, g, A) > 1$ such that, if $v > 0$ solves*

$$L_g v = K v^{\frac{n+2}{n-2}-\tau}, \quad \max_M v > C_0, \quad \lambda_1(L_g) > 0,$$

then there exists an integer $L = L(v)$ and a finite set of points $S(v) = \{q_1, \dots, q_L\}$ such that

(1) *$\tau < \varepsilon$;*

(2) *$q_1, \dots, q_L$ are local maxima of v and (in geodesic coordinates at q_j)*

$$\|v(0)^{-1} v(v(0)^{-\frac{p-1}{2}} x) - C_{K(q_j)} U(x)\|_{C^2(B_{2R}(0))} < \varepsilon,$$

and the balls $(B_{Rv(q_j)^{-(p-1)/2}}(q_j))_j$ are disjoint for $j = 1, \dots, L(v)$, where $p = \frac{n+2}{n-2} - \tau$ and $C_{K(q_i)}$ is as in Corollary 6.8;

(3) *$v(x) < C_1 \mathrm{dist}(x, S(v))^{-2/(p-1)}$.*

The proof is by induction and based on the following lemma.

Lemma 6.13. *Let K be as in Proposition 6.12. Let $\widetilde{M}$ be a compact subset of M. Then for every $\varepsilon > 0$ and $R \gg 1$ there exists a constant $C = C(\varepsilon, R, M, g, A)$ such that if v is a solution to (we set $\mathrm{dist}(x, \widetilde{M}) = 1$ for $\widetilde{M} = \emptyset$)*

$$L_g u = u^p \ \text{ on } M \setminus \widetilde{M}, \quad \max_{M \setminus \widetilde{M}} \mathrm{dist}(x, \widetilde{M})^{\frac{2}{p-1}} u(x) \geq C,$$

for some $p \in [1, \frac{n+2}{n-2}]$, then $\frac{n+2}{n-2} - p < \varepsilon$ and there exists $x_0 \in M \setminus \widetilde{M}$ such that x_0 is a local maximum of u, in geodesic coordinates at x_0,

$$\|u(0)^{-1} u(u(0)^{-\frac{p-1}{2}} x) - C_{K(x_0)} U(x)\|_{C^2(B_{2R}(0))} < \varepsilon,$$

and $B_{Ru(0)^{-(p-1)/2}}(0) \subseteq M \setminus \widetilde{M}$.

Proof. Assume by contradiction that no such constant C exists. Then there are $(u_i)_i$, $(K_i)_i$, $(p_i)_i$, and $\widetilde{M}_i$ such that

$$\max_{M \setminus \widetilde{M}_i} d(x, \widetilde{M}_i)^{\frac{2}{p_i - 1}} u_i(x) \geq 1,$$

and for each i, no such point x_0 exists. Set

$$f_i(x) := d(x, \widetilde{M}_i)^{\frac{2}{p_i-1}} u_i(x), \quad f_i(x_i) := \max_{M \setminus \widetilde{M}_i} f_i.$$

By Theorem 5.8 one can show that necessarily $p_i \to \frac{n+2}{n-2}$. Define

$$w_i(x) := u_i(x_i)^{-1} u_i(u_i(x_i)^{-\frac{p_i-1}{2}} x), \quad R_i := \tfrac{1}{2} d(x_i, \widetilde{M}_i) u_i(x_i)^{\frac{p_i-1}{2}},$$

where x are the normal coordinates centered at x_i. Then

$$w_i(0) = 1, \quad \sup_{B_{R_i}} w_i \leq C.$$

Hence, by standard elliptic theory, $w_i \to w_\infty$ in $C^2_{\mathrm{loc}}(\mathbb{R}^n)$, with w_∞ satisfying

$$-\Delta w_\infty = k_0 w_\infty^{\frac{n+2}{n-2}} \quad \text{on } \mathbb{R}^n,$$

for some $k_0 := K(x_\infty) \in [\frac{1}{A}, A]$. By Theorem 5.1, w_∞ is of the form $C(K(x_\infty))U_{\lambda,\bar{x}}$, for some $\bar{x} \in \mathbb{R}^n$, $\lambda > 0$. This shows that the maximum of w_i is attained within a fixed ball (with radius independent of i) centered at the origin. Hence, up to a small translation and dilation, we see that

$$\| w_i - C(K(x_\infty))U \|_{C^2(B_{3R})} \to 0,$$

a contradiction to our assumption. ∎

Proof of Proposition 6.12. Let q_1 be a maximum point of v, and then set $r_1 := Rv(q_1)^{-2/(p-1)}$. Then, applying the above lemma with $\widetilde{M} = \emptyset$, we see that property (2) holds. Now let $\widetilde{M}_1 := \overline{B_{r_1}(q_1)}$. If $d(x, \widetilde{M}_1)^{2/(p-1)} v(x) \leq C$, then property (3) is satisfied, and we are done. Otherwise, we apply the above lemma again to find $q_2 \in M \setminus \widetilde{M}_1$ such that property (2) holds with q_1 and q_2. Repeating these arguments we see that the process ends after a finite number of steps as M is closed and the radii r_i satisfy $r_i \geq R(\max u)^{-2/(p-1)}$. ∎

Proposition 6.14. *Let $(K_i)_i$, $(u_i)_i$ be as in Theorem* 6.9. *Then blow-ups are isolated.*

Proof. Let $L_i := L(u_i)$ be as in Proposition 6.12, and let $\sigma_i := \min_{j \neq \ell} \mathrm{dist}_g(q_j, q_\ell)$. Without loss of generality we can assume that $\sigma_i = \mathrm{dist}(q_1, q_2)$. If by contradiction $\sigma_i \to 0$, use geodesic coordinates at q_1 and set

$$w_i(x) := \sigma_i^{\frac{2}{p_i-1}} u_i(\sigma_i x), \quad \bar{q} := \lim_{i \to \infty} \frac{q_{2,i}}{\sigma_i}, \quad \bar{q}_i := \frac{q_{2,i}}{\sigma_i}.$$

It follows that $|\bar{q}| = 1$. By Proposition 6.12 (2),

$$\sigma_i \gg \max\{u_i(q_1)^{-\frac{p_i-1}{2}}, u_i(q_2)^{-\frac{p_i-1}{2}}\},$$

and hence w_i blows up at 0 and $\bar{q}$. Moreover, 0 and $\bar{q}$ are isolated, and isolated simple by Proposition 6.10. Using Proposition 6.7 and Harnack's inequality we see that there exists a discrete set $S_1 \subseteq \mathbb{R}^n$ containing 0 and $\bar{q}$ such that

$$w_i(0)w_i(x) \to h(x) \text{ in } C^2_{\mathrm{loc}}(\mathbb{R}^n \setminus S_1), \quad h > 0,\ \Delta h = 0 \text{ on } \mathbb{R}^n \setminus S_1.$$

Moreover, h is singular at all points of S_1 (at 0 and $\bar{q}$, in particular). Near 0 one has

$$h(x) = a|x|^{2-n} + b(x) + O(x).$$

Since $h - a|x|^{2-n}$ cannot attain a non-positive minimum, we have that $b(\cdot) > 0$. We now get a contradiction as for the proof of Proposition 6.10. ■

Proof of Theorem 6.9. From Propositions 6.7, 6.10, 6.12, and 6.14 we have that all blow-ups are isolated simple, and therefore we have the validity of the formula in Remark 6.11 (i).

If $n = 3$, we must then have

$$\alpha_1 A(\bar{x}) + \sum_{j=2}^{L} \alpha_j G_{\bar{x}^j}(\bar{x}) = 0.$$

Since $A(\bar{x}) \geq 0$ by the positive mass theorem (see Remark 3.15) and since $\alpha_j > 0$, there can be at most one blow-up point, at which $\Delta_g K < 0$. If (M^3, g) is not conformally equivalent to $(\mathbb{S}^3, g_{\mathbb{S}^3})$, then blow-up cannot occur since $A(\bar{x}) > 0$.

For $n = 4$, the formula in Remark 6.11 (i) implies that, even in the case of multiple blow-up points, at each of them the Laplacian of K must be negative since, again, $A \geq 0$ and $\alpha_j > 0$ for all j.

Recall also that, by Remark 6.11 (ii), blow-ups only occur at critical points of K. ■

As a corollary of the latter result we also obtain the following theorem.

Theorem 6.15 ([157]). *If (M^3, g) is of positive Yamabe class and it is not conformally equivalent to $(\mathbb{S}^3, g_{\mathbb{S}^3})$, then (E_K) is solvable for any positive (regular) function K.*

Proof. As we remarked at the beginning of the chapter, there always exist (positive) solutions of $(E_{K,\tau})$, and by Theorem 6.9 they must stay uniformly bounded. On the other hand, as one can see by multiplying $(E_{K,\tau})$ by u, integrating by parts, and using the Sobolev inequality, solutions of $(E_{K,\tau})$ cannot converge to zero either, so they must possess a positive limit, solving (E_K). ■

Remark 6.16. We notice the following facts:

(i) For $n = 4$, blow-up at multiple points is possible, but with some constraints on their location due to formula (6.12); see [29, 151]. Assuming that K is Morse, positive, and that

$$\frac{\Delta K(x^i)}{3K(x^i)} + 2\frac{A(x^i)}{K(x^i)} \neq 0$$

at all critical points x^i, define the matrix

$$M_{ij} := \begin{cases} -\frac{\Delta K(x^i)}{3K(x^i)} - 2\frac{A(x^i)}{K(x^i)} & \text{for } i = j, \\ -\frac{2G_{x^i}(x^j)}{(K(x^i)K(x^j))^{\frac{1}{2}}} & \text{for } i \neq j. \end{cases}$$

Multi-bubbling can only occur if the least eigenvalue of $(M_{ij})_{ij}$ is positive.

(ii) Conclusions similar to those in Theorem 6.9 hold true in all dimensions if K_i, K are *flat* of order $\beta \geq n - 2$; see [158]. However, as shown in [148], such conclusions are false in general, since solutions forming collapsing multiple bubbles can be constructed; see also Remark 9.1.

(iii) For $n \geq 5$ and K a Morse function, a more detailed blow-up analysis was carried out in [62,63]. It is shown there that, under some general circumstances, blow-ups can be still isolated. On the other hand, sequences of blowing-up solutions might sometimes have non-zero weak limits and may display concentration of infinite volume. Some of this analysis will indeed be employed in Chapter 9, and more details will be presented in Chapter 10.

(iv) On $(\mathbb{S}^2, g_{\mathbb{S}^2})$ there can be at most one blow-up point for the prescribed Gaussian curvature problem for arbitrary positive functions, as proven in [57].

(v) A case that needs special attention is that of the Yamabe problem for $\lambda_1(L_g) > 0$, namely for K identically equal to a positive constant. After the contribution in [157] mentioned in Theorem 6.15, compactness of solutions to the Yamabe problem on manifolds not conformally equivalent to $(\mathbb{S}^n, g_{\mathbb{S}^n})$ has been proved in [96, 145, 159] via a more refined blow-up analysis up to dimension 24. On the other hand, after some first examples in [31] for metrics of class C^k (see also [7]), sequences of diverging solutions on manifolds with C^∞-metrics were constructed in [43] for dimension $n \geq 25$. The surprising dimensional threshold depends on the eigenvalues of a quadratic form that plays a role in the Pohozaev identity on general Riemannian manifolds.

Chapter 7

Finite-dimensional reduction of $(\mathrm{E}_{K,\tau})$

Let (M, g) be closed of positive Yamabe class, i.e. such that $\lambda_1(L_g) > 0$. Let $K > 0$ be a Morse function satisfying the non-degeneracy condition (ND). We recall from the previous section (see Theorem 6.9) that blow-ups for problem $(\mathrm{E}_{K,\tau})$ are isolated simple in dimensions $n = 3, 4$, and that they occur at critical points of K with negative Laplacian.

We also recall equation (6.12) from Remark 6.11: if we are on $(\mathbb{S}^n, g_{\mathbb{S}^n})$ and in the case of a single blow-up point, this formula implies that (up to some positive coefficients in each factor)

$$\Delta K(\bar{x})\lambda_i^{-2} + \tau \simeq 0.$$

As a consequence, from the condition $\Delta K(\bar{x}^1) < 0$ we obtain

$$\tau \approx \lambda_i^{-2} \approx u_i(x_i)^{-\frac{4}{n-2}}.$$

We will show in this chapter that there are indeed solutions blowing up at this rate, namely that the following theorem holds true.

Theorem 7.1 ([25, 151, 158, 164]). *Let $K\colon \mathbb{S}^n \to \mathbb{R}$ be a positive Morse function, and let $\bar{x}$ be a non-degenerate critical point of K with $\Delta K(\bar{x}) < 0$. Then there exists $u_{\tau,\bar{x}}$ solution of $(\mathrm{E}_{K,\tau})$ such that $\bar{x}$ is an isolated-simple blow-up. Such a solution is unique and satisfies the following properties:*

$$\tau \approx (\max_{\mathbb{S}^n} u_{\tau,\bar{x}})^{-\frac{4}{n-2}},$$

$$\operatorname{ind}(J_{K,\iota}, u_{\tau,\bar{x}}) = n - \operatorname{ind}(K, \bar{x}),$$

where ind *denotes the Morse index.*

Recall that minima of $J_{K,\tau}$ for τ small concentrate near maxima of K; see the beginning of Chapter 6. More generally, each stable direction for a critical point of K generates an unstable direction for the corresponding critical point of J_K.

Remark 7.2. In further generality, we have the following results:

(i) If $n = 4$, blow-up at multiple points is possible, but with some constraints on their location. On a closed manifold (M^4, g) of positive Yamabe class, let K be a Morse function. Given a subset of its critical points $(x^i)_i$, consider the matrix M_{ij} given in Remark 6.16, namely

$$M_{ij} := \begin{cases} -\dfrac{\Delta K(x^i)}{3K(x^i)} - 2\dfrac{A(x^i)}{K(x^i)} & \text{for } i = j, \\ -\dfrac{2G_{x^i}(x^j)}{(K(x^i)K(x^j))^{\frac{1}{2}}} & \text{for } i \neq j. \end{cases}$$

Assuming $M_{ii} \neq 0$ for every critical point x^i of K, then one can construct solutions as in the previous theorem when the least eigenvalue of $(M_{ij})_{ij}$ is positive; see [29, 151, 164]. On Euclidean domains, the role of the Green's function in constructions of this kind was also shown in [82, 190].

(ii) If (M^n, g) is an arbitrary closed manifold of positive Yamabe class, $n \geq 5$ and $K > 0$, is Morse, and satisfies (ND), given any q-tuple $\{x^1, \dots, x^q\}$ of critical points of K with $\Delta K(x^i) < 0$ for all i, there exists $u_{\tau,x^1,\dots,x^q}$ blowing up at each x^i. Contrary to the four-dimensional case, there is no restriction on the blow-up set. The Morse index of this solution is

$$(q-1) + \sum_{i=1}^{q} (n - \operatorname{ind}(K, x^i)).$$

Similarly to Theorem 7.1, each stable direction of K yields an instability direction for $u_{\tau,x^1,\dots,x^q}$; moreover, extra instabilities are given by the presence of multiple bubbles, and are generated from their multiplication by different constants.

(iii) As for Theorem 7.1, the solutions described in items (i) and (ii) of this remark are unique.

For the standard three-sphere, we have in particular the following complete characterization of the blow-up behavior, which will be useful to prove Theorem C. It follows by combining Theorem 7.1 with Theorem 4.6.

Corollary 7.3. *On $\mathbb{S}^3$, blowing-up solutions of $(E_{K,\tau})$ as $\tau \to 0$ are in one-to-one correspondence with $\{\nabla K = 0\} \cap \{\Delta K < 0\}$.*

7.1 Lyapunov–Schmidt reduction

Theorem 7.1 can be proved via a Lyapunov–Schmidt reduction of the problem, jointly with the variational structure of $(E_{K,\tau})$. We refer to the book [9] for a general treatment of such methods, where different applications are treated.

Recall the expression of the spherical bubble $\varphi_{p,\lambda}$ from (3.3), and define

$$Z_0 = \{\alpha\varphi_{p,\lambda} : p \in \mathbb{S}^n,\ \lambda > 0,\ \alpha > 0\}.$$

In Chapter 3 it was proved that Z_0 forms a manifold of minima for the spherical Yamabe quotient $Q_{g_{\mathbb{S}^n}}$ on $\mathbb{S}^n$. We show next that this manifold is indeed minimally degenerate.

Proposition 7.4. *The manifold Z_0 is non-degenerate for $Q_{g_{\mathbb{S}^n}}$ in the sense of Bott, namely the second differential of $Q_{g_{\mathbb{S}^n}}$ is positive definite orthogonally to Z_0 (see Figure 7.1).*

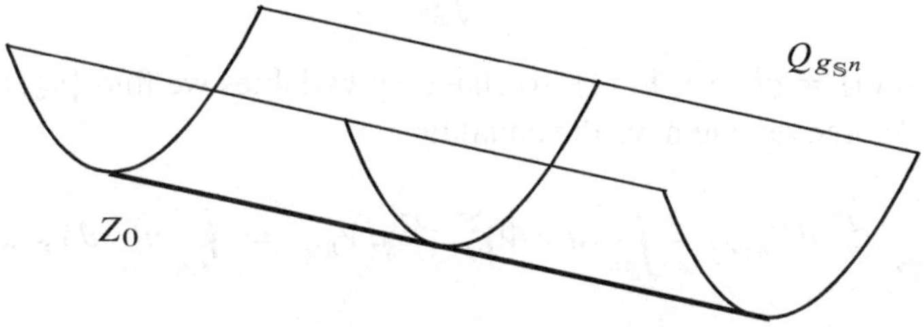

Figure 7.1. Non-degeneracy of $Q_{g_{\mathbb{S}^n}}$ in the sense of Bott

Proof. It is sufficient to show that for every $z \in Z_0$, $\operatorname{Ker} d^2(Q_{g_{\mathbb{S}^n}})(z) = T_z Z_0$. Since the inclusion $T_z Z_0 \subseteq \operatorname{Ker} d^2(Q_{g_{\mathbb{S}^n}})(z)$ is trivially obtained differentiating the Euler–Lagrange equation with respect to the parameters defining Z_0, it will be sufficient to prove the opposite one. We proceed in two steps.

Step 1 (Verify that $\operatorname{Ker} d^2(Q_{g_{\mathbb{S}^n}})(z) \subseteq T_z Z_0$ for $Z_0 \ni z \equiv 1$). As we saw, the criticality condition $dQ_{g_{\mathbb{S}^n}}(u) = 0$ comes with a Lagrange multiplier. Since $Q_{g_{\mathbb{S}^n}}$ is homogeneous of degree zero, we can always scale its argument so that the stationarity equation becomes

$$L_g u = n(n-1)u^{\frac{n+2}{n-2}} \quad \text{on } \mathbb{S}^n. \tag{7.1}$$

We need therefore to classify solutions of the linearized equation

$$L_g v - n(n-1)\frac{n+2}{n-2} v = 0 \quad \text{on } \mathbb{S}^n,$$

which simply means

$$-\Delta_g v = nv.$$

The solutions of this eigenvalue equation are the linear functions of $\mathbb{R}^{n+1}$ restricted to $\mathbb{S}^n$, an $(n+1)$-dimensional space spanning $T_1 Z_0$, up to the constant functions on $\mathbb{S}^n$.

Since all other eigenvalues of $-\Delta_{g_{\mathbb{S}^n}}$ on $\mathbb{S}^n$ are larger than n (see e.g. [9, Chapter 4]), the second differential of $Q_{g_{\mathbb{S}^n}}$ at $z \equiv 1$ is positive definite orthogonally to Z_0.

Step 2 (Case of general $z \in Z_0$). As in the previous step, we can assume that z is normalized so that it satisfies (7.1).

Under this condition, there exists a Möbius map $\Phi\colon \mathbb{S}^n \to \mathbb{S}^n$ such that $\tilde{g} := \Phi^* g_{\mathbb{S}^n} = z^{\frac{4}{n-2}} g_{\mathbb{S}^n}$. The map $u(x) \mapsto u_z(x) := z(x)u(\Phi(x))$ gives an isometry of $H^1(\mathbb{S}^n)$ in itself: in fact, by (2.2),

$$\begin{aligned}\langle u_z, v_z\rangle_{H^1(\mathbb{S}^n)} = \int_{\mathbb{S}^n} u_z L_{g_{\mathbb{S}^n}} v_z \, dV_{g_{\mathbb{S}^n}} &= \int_{\mathbb{S}^n} u(\Phi(x)) z(x) L_{g_{\mathbb{S}^n}} z(x) v(\Phi(x)) \, dV_{g_{\mathbb{S}^n}} \\ &= \int_{\mathbb{S}^n} u(\Phi(x)) L_{\tilde{g}}(v \circ \Phi(x)) z^{2^*}(x) \, dV_{g_{\mathbb{S}^n}}.\end{aligned}$$

But $L_{\tilde{g}}(v \circ \Phi(x)) = (L_g v)(\Phi(x))$, so changing variables we find $\langle u_z, v_z\rangle_{H^1(\mathbb{S}^n)} = \langle u, v\rangle_{H^1(\mathbb{S}^n)}$. Moreover, we have the equalities

$$\int_{\mathbb{S}^n} u_z^{2^*} \, dV_{g_{\mathbb{S}^n}} = \int_{\mathbb{S}^n} (u \circ \Phi)^{2^*} z^{2^*} \, dV_{g_{\mathbb{S}^n}} = \int_{\mathbb{S}^n} u^{2^*} \, dV_{g_{\mathbb{S}^n}}.$$

Hence, $Q_{g_{\mathbb{S}^n}}(u) = Q_{g_{\mathbb{S}^n}}(u_z)$ for every $u \in H^1(\mathbb{S}^n)$. Therefore, still assuming that the Lagrange multiplier is normalized as above, we deduce the relation

$$\operatorname{Ker} d^2 Q_{g_{\mathbb{S}^n}}(z) = (\operatorname{Ker} d^2 Q_{g_{\mathbb{S}^n}}(1))_z,$$

with $\operatorname{Ker} d^2 Q(1)$ as in Step 1. We then conclude by noticing that $(T_1 Z_0)_z = T_z Z_0$. ∎

As we will see, for $\lambda \simeq \tau^{-\frac{1}{2}}$ the functions $(\frac{K(p)}{n(n-1)})^{\frac{4}{n-2}} \varphi_{p,\lambda}$ are approximate solutions of $(\mathrm{E}_{K,\tau})$. We will show how to turn them into true solutions via a general abstract approach: we refer to the book [9] for more details.

Setting. Let X be a Hilbert space, $J_\tau\colon X \to \mathbb{R}$ a family of C^2-functionals, $Z_\tau \subseteq X$ finite-dimensional sub-manifolds (of dimension independent of τ) such that the following properties hold true:

(A) $\|dJ_\tau(z)\| \le A_\tau$ for every $z \in Z_\tau$, $A_\tau \to 0$ as $\tau \to 0$.

(B) There exist $\Gamma, \delta > 0$, $\alpha \in (0,1)$ such that $\|J_\tau\|_{C^{2,\alpha}(B_\delta(z))} \le \Gamma$ for every $z \in Z_\tau$.

(C) There exists $\beta > 0$ such that $\|d^2 J_\tau(z)[v,\cdot]\| \ge \beta \|v\|$ for every $v \perp T_z Z_\tau$.

(D) $\|d^2 J_\tau(z)[v,\cdot]\| \le B_\tau$ for every $z \in Z_\tau$ and every $v \in T_z Z$, with $B_\tau \to 0$ as $\tau \to 0$.

Hypothesis (C) implies that we have a good control on the component of ∇J_τ normal to Z_τ when moving normally to Z_τ. This fact will help to prove the next proposition.

Proposition 7.5. *Suppose that* (A), (B), *and* (C) *hold. Then for every $z \in Z_\tau$ there exists a (locally) unique $w = w_\tau(z)$ such that*

(i) $w_\tau(z) \perp T_z Z_\tau$;

(ii) $\nabla J_\tau(z + w)$ *is parallel to* $T_z Z_\tau$.

Moreover, there exists a positive constant $\bar{C}$ such that $\|w\| \leq \bar{C} A_\tau$.

Proof. The proof uses a contraction argument. We let

$$P_z \colon X \to W_z := (T_z Z_\tau)^\perp$$

denote the orthogonal projection, and our aim is to find $w \in W_z$ such that $P_z \nabla J_\tau(z + w) = 0$. By property (C) we know that $P_z d^2 J_\tau(z)$ is uniformly invertible, and therefore we can write $P_z \nabla J_\tau(z + w) = 0$ as

$$0 = P_z \nabla J_\tau(z + w) + P_z \nabla J_\tau(z) - P_z \nabla J_\tau(z) + P_z \nabla^2 J_\tau(z)[w] - P_z \nabla^2 J_\tau(z)[w],$$

which by the above invertibility is equivalent to

$$w = S_z(w) := -P_z \nabla^2 J_\tau(z)^{-1}\big[P_z \nabla J_\tau(z) + R_z(w)\big],$$

where

$$R_z(w) = P_z \nabla J_\tau(z + w) - P_z \nabla J_\tau(z) - \nabla^2 J_\tau(z)[w].$$

We next want to prove that S_z is a contraction from a suitable norm ball in W_z to itself. In fact, given $C > 0$, using (B) for $w_1, w_2 \in B_{CA_\tau}(z)$ we first have the estimate

$$R_z(w_1) - R_z(w_2) = \int_0^1 \big[\nabla^2 J_\tau(z + w_1 + s(w_2 - w_1)) - \nabla^2 J_\tau(z)\big][w_1 - w_2]\, ds,$$

which by (B) yields

$$\|R_z(w_1) - R_z(w_2)\| \leq C_\Gamma \|w_1 - w_2\|^{1+\alpha}.$$

This fact, together with properties (C) and (A), then implies

$$\|S_z(w)\| \leq \bar{C}(A_\tau + C_\Gamma \|w\|^{1+\alpha}),$$

and that S_z is a contraction in a norm ball in W_z of radius CA_τ. This concludes the proof. ∎

Corollary 7.6. *Assume that* (D) *also holds true, and if* $w_\tau(z)$ *is as in the previous proposition, define the perturbed manifold*

$$\widetilde{Z}_\tau := \{z + w_\tau(z) \colon z \in Z_\tau\}.$$

Then $\widetilde{Z}_\tau$ *is a natural constraint for* J_τ*, i.e. if* $\tilde{z}$ *is critical for* $J_\tau|_{\widetilde{Z}_\tau}$*, then* $\tilde{z}$ *is also critical for* J_τ*.*

Proof. Recall that $w_\tau(z)$ is found as a fixed point for the above contraction mapping S_z. When a contraction depends on some parameters, z in this case, the fixed point has the same smoothness as $z \mapsto S_z$; see [44, Section 2.6]. Using condition (D) it is possible to prove that $\partial_z w_\tau(z)$ has small norm as $\tau \to 0$. This implies that $T_z Z_\tau$ is nearly parallel to $T_{z+w}\widetilde{Z}$.

By the Lagrange multiplier rule $\nabla J(\tilde{z}) \perp T_{\tilde{z}}\widetilde{Z}_\tau$, but by construction $\nabla J(z)$ is parallel to $T_z Z_\tau$. Since $T_{\tilde{z}}\widetilde{Z}_\tau \simeq T_z Z_\tau$, it must be $\nabla J_\tau(\tilde{z}) = 0$ (see Figure 7.2 (ii)). ∎

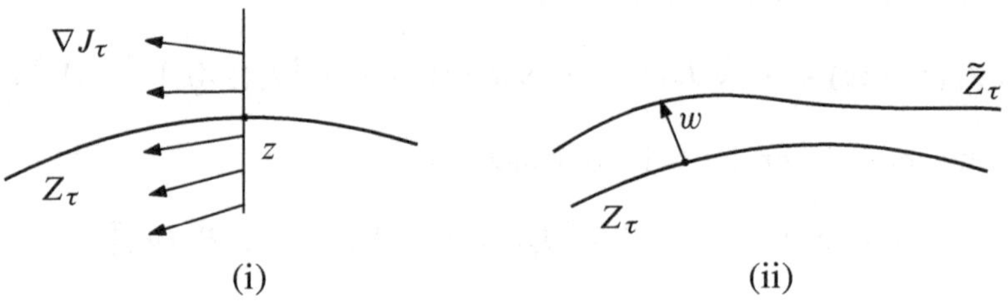

Figure 7.2. Normal variation of the gradient and the natural constraint

By the latter corollary, it is useful to find critical points of $J_\tau|_{\widetilde{Z}_\tau}$: this can be done by Taylor-expanding J_τ around z, for $z \in Z_\tau$.

Proposition 7.7. *Define* $F_\tau \colon Z_\tau \to \mathbb{R}$ *by*

$$F_\tau(z) := J_\tau(z + w_\tau(z)).$$

Then

$$F_\tau(z) = J_\tau(z) + O(A_\tau^2) \quad \text{for every } z \in Z_\tau.$$

Proof. We have

$$F_\tau(z) = J_\tau(z) + dJ_\tau(z)[w_\tau(z)] + O(\|w\|^2).$$

Now use that $\|\nabla J_\tau(z)\| \lesssim A_\tau$ and $\|w_\tau(z)\| \lesssim A_\tau$ to obtain the conclusion. ∎

7.2 Existence of blowing-up solutions to $(E_{K,\tau})$

Our aim is to apply Corollary 7.6 to the specific case of $J_\tau = J_{K,\tau}$ as in (6.1), for a suitable choice of Z_τ.

Proof of Theorem 7.1. Let $X = H^1(\mathbb{S}^n)$, $J_\tau = J_{K,\tau}$, and given a large constant C define

$$Z_\tau := \{\varphi_{q,\lambda} \in Z_0 \colon \lambda \in [C^{-1}\tau^{-\frac{1}{2}}, C\tau^{-\frac{1}{2}}],\ d(q,\bar{x}) \le C\tau\}.$$

With the aim of applying Proposition 7.5, we notice that property (B) is standard, while (C) follows from Proposition 7.4. Next we state two lemmas, whose proofs are given below.

Lemma 7.8. *If q is as in the above definition of Z_τ,* (A) *holds with*

$$A_\tau = O_C(\tau|\log\tau|).$$

Lemma 7.9. *For $\lambda \in [C^{-1}\tau^{-\frac{1}{2}}, C\tau^{-\frac{1}{2}}]$ one has the expansion*

$$J_{K,\tau}(\varphi_{q,\lambda}) \approx \lambda^\tau(K(q)^{-1} - \lambda^{-2}\Delta K(q)), \tag{7.2}$$

and (see Figure 7.3 (i)*)*

$$\min\{J_{K,\tau}(\varphi_{q,C^{-1}\tau^{-\frac{1}{2}}}), J_{K,\tau}(\varphi_{q,C\tau^{-\frac{1}{2}}})\} \gtrsim J_{K,\tau}(\varphi_{q,\tau^{-\frac{1}{2}}}) + \tau\log C.$$

Proof of Theorem 7.1 *continued.* By Proposition 7.7 and Lemma 7.8, the function $\lambda \mapsto F_\tau(\varphi_{q,\lambda})$ has an interior minimum λ_τ in the interval $[C^{-1}\tau^{-\frac{1}{2}}, C\tau^{-\frac{1}{2}}]$. From Lemma 7.9 it is not hard to show that there is also a critical point q_τ of $q \mapsto F_\tau(\varphi_{q,\lambda})$ with $q_\tau \to \bar{x}$ as $\tau \to 0$. As property (D) can be proved similarly to (A), we can then

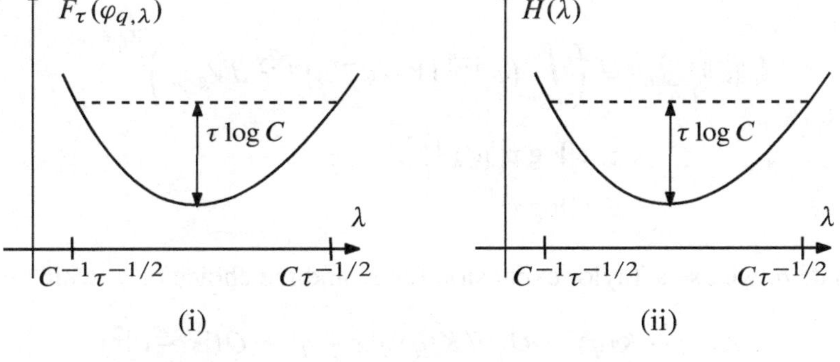

Figure 7.3. Oscillation of the reduced functional

apply Corollary 7.6 to deduce existence of a critical point close in $H^1(\mathbb{S}^n)$ to $\varphi_{q_\tau,\lambda_\tau}$. Standard elliptic regularity results imply that the solution constructed in this way develops an isolated simple blow-up.

Further Taylor expansions of $F_\tau(\varphi_{q,\lambda})$, not performed here, show that $u_{\tau,\bar{x}}$ is non-degenerate, with Morse index for $J_{K,\tau}$ given as in the statement of the theorem (coinciding with that of the critical point in (λ,τ) for the right-hand side of (7.2)). We refer to [164] for complete details. ∎

Proof of Lemma 7.8. For any $v \in H^1(\mathbb{S}^n)$ we have

$$\begin{aligned}\langle \nabla J_{K,\tau}(\varphi_{q,\lambda}), v\rangle &:= dJ_{K,\tau}(\varphi_{q,\lambda})[v] \\ &= 2\Theta^{-\frac{2}{p-1}}\bigg[\int_{\mathbb{S}^n} v L_{g_{\mathbb{S}^n}}\varphi_{q,\lambda}\, dV_{g_{\mathbb{S}^n}} \\ &\qquad - \Big(\int_{\mathbb{S}^n}\varphi_{q,\lambda}L_{g_{\mathbb{S}^n}}\varphi_{q,\lambda}\, dV_{g_{\mathbb{S}^n}}\Big)\Theta^{-1}\int_{\mathbb{S}^n} K\varphi_{q,\lambda}^p v\, dV_{g_{\mathbb{S}^n}}\bigg],\end{aligned}$$

where $p = \frac{n+2}{n-2} - \tau$ and $\Theta = \int_{\mathbb{S}^n} K\varphi_{q,\lambda}^{2^*-\tau}\, dV_{g_{\mathbb{S}^n}}$.

Let us now set

$$a := K(q)\big(\varphi_{q,\lambda}^{\frac{n+2}{n-2}} - \varphi_{q,\lambda}^p\big), \quad b := (K(q)-K)\varphi_{q,\lambda}^p,$$

and show next that

$$\|a\|_{L^{\frac{2n}{n+2}}} + \|b\|_{L^{\frac{2n}{n+2}}} \lesssim \tau|\log\tau|.$$

To prove the estimate for a, we recall that $\max\varphi_\lambda \approx \lambda^{\frac{n-2}{2}}$ (here and below we simply write φ_λ for $\varphi_{q,\lambda}$), and that $\lambda \approx \tau^{-\frac{1}{2}}$. Therefore,

$$\|1-\varphi_\lambda^{-\tau}\|_{L^\infty} \le \Big|1-\exp\Big(\frac{n-2}{2}\tau\log\lambda\Big)\Big| \lesssim \tau|\log\tau|.$$

This gives

$$\begin{aligned}\|a\|_{L^{\frac{2n}{n+2}}} &= \Big(\int_{\mathbb{S}^n}\big(\varphi_\lambda^{\frac{n+2}{n-2}}(1-\varphi_\lambda^{-\tau})\big)^{\frac{2n}{n+2}}\, dV_{g_{\mathbb{S}^n}}\Big)^{\frac{n+2}{2n}} \\ &\lesssim \tau|\log\tau|\,\|\varphi_\lambda\|_{L^{2^*}}^{\frac{n+2}{n-2}} \\ &\lesssim \tau|\log\tau|.\end{aligned}$$

To estimate b, we use a Taylor expansion for K and the choice of q, which gives

$$\begin{aligned}K(x)-K(q) &= O(|\nabla K(q)|)|x-q| + O(|x-q|^2) \\ &= O(\tau|x-q|) + O(|x-q|^2).\end{aligned}$$

Using a change of variables and the fact that φ_λ conformally corresponds to U_λ in stereographic coordinates, this gives

$$\begin{aligned}\|b\|_{L^{\frac{2n}{n+2}}} &\lesssim \left(\int_{\mathbb{R}^n} |U_\lambda|^{2^*} \left(|x|^{2\frac{2n}{n+2}} + \tau^{\frac{2n}{n+2}}|x|^{\frac{2n}{n+2}}\right) dx\right)^{\frac{n+2}{2n}} \\ &\lesssim \left(\lambda^{-\frac{4n}{n+2}} + \tau^{\frac{2n}{n+2}}\lambda^{-\frac{2n}{n+2}}\right)^{\frac{n+2}{2n}} \\ &\simeq \lambda^{-2} \simeq \tau,\end{aligned}$$

which yields the desired claim.

Using the fact that $L_{g_{\mathbb{S}^n}}\varphi_{q,\lambda} = K(q)\varphi_{q,\lambda}^{\frac{n+2}{n-2}}$, we find that

$$\begin{aligned}|\Theta - L_{g_{\mathbb{S}^n}}\varphi_{q,\lambda}| &= \left|\int_{\mathbb{S}^n} (a+b)\varphi_{q,\lambda}\right| \\ &\le (\|a\|_{L^{\frac{2n}{n+2}}} + \|b\|_{L^{\frac{2n}{n+2}}})\|\varphi_{q,\lambda}\|_{L^{2^*}} \le C\tau|\log\tau|.\end{aligned}$$

From the initial formula in the proof one then has that

$$\begin{aligned}dJ_{K,\tau}(\varphi_{q,\lambda})[v] &= 2\Theta^{-\frac{2}{p-1}}\left[\int_{\mathbb{S}^n} vL_{g_{\mathbb{S}^n}}\varphi_{q,\lambda}\, dV_{g_{\mathbb{S}^n}} - \int_{\mathbb{S}^n} K\varphi_{q,\lambda}^p v\, dV_{g_{\mathbb{S}^n}}\right] \\ &\quad + O(\tau|\log\tau|)\|v\|_{H^1(\mathbb{S}^n)} \\ &= \int_{\mathbb{S}^n} (a+b)\varphi_{q,\lambda} v\, dV_{g_{\mathbb{S}^n}}.\end{aligned}$$

Using Hölder's inequality again we then find

$$\begin{aligned}|dJ_{K,\tau}(\varphi_{q,\lambda})[v]| &\le C(\|a\|_{L^{2^*}} + \|b\|_{L^{2^*}})\|v\|_{L^{2^*}} + C\tau|\log\tau|\|v\|_{L^{2^*}} \\ &\le C\tau|\log\tau|\|v\|_{H^1}.\end{aligned}$$

By the arbitrariness of v, we obtain the conclusion. ∎

Proof of Lemma 7.9. Recall that the numerator in $J_{K,\tau}(\varphi_\lambda)$ is the same as that of $Q_{g_{\mathbb{S}^n}}(\varphi_\lambda)$, and hence it is independent of λ; see Remark 3.10. So we focus on the term $\int_{\mathbb{S}^n} K\varphi_\lambda^{2^*-\tau}\, dV_{g_{\mathbb{S}^n}}$ in the denominator.

In the formulas below, for simplicity, we will not be careful with multiplicative constants, with the goal of showing the general dependence of the expansions in λ and τ. Assuming this is understood, Taylor-expanding K we obtain

$$\int_{\mathbb{S}^n} K\varphi_{q,\lambda}^{2^*-\tau}\, dV_{g_{\mathbb{S}^n}} \approx K(q)\lambda^{-\tau} + \Delta K(q)\lambda^{-2-\tau},$$

and hence (replacing the exponent $-\frac{2}{2^*-\tau}$ with -1)

$$\left(\int_{\mathbb{S}^n} K\varphi_\lambda^{2^*-\tau}\, dV_{g_{\mathbb{S}^n}}\right)^{-1} \approx \lambda^\tau(K(q)^{-1} - \lambda^{-2}\Delta K(q)).$$

For $\lambda \in [C^{-1}\tau^{-\frac{1}{2}}, C\tau^{-\frac{1}{2}}]$, $K(q) = 1$, $\Delta K(q) = -1$ we have

$$\lambda^{\tau}(K(q)^{-1} - \lambda^{-2}\Delta K(q)) \approx \lambda^{\tau}(1 + \lambda^{-2}) \approx 1.$$

To estimate the oscillation in λ of the latter quantity we can take the logarithm, which is

$$H(\lambda) := \tau \log \lambda + \log(1 + \lambda^{-2}) \approx \tau \log \lambda + \lambda^{-2}.$$

We have

$$\begin{aligned} H(C^{-1}\tau^{-\frac{1}{2}}) &\approx \tau(-\tfrac{1}{2}\log\tau - \log C) + C^2\tau \approx -\tfrac{1}{2}\tau\log\tau + C^2\tau, \\ H(\tau^{-\frac{1}{2}}) &\approx -\tfrac{1}{2}\tau\log\tau, \\ H(C\tau^{-\frac{1}{2}}) &\approx \tau(-\tfrac{1}{2}\log\tau + \log C) + C^{-2}\tau \approx -\tfrac{1}{2}\tau\log\tau + \tau\log C. \end{aligned}$$

This is the desired behavior; see Figure 7.3 (ii). ■

Remark 7.10. We notice the following facts:

(i) The map S_z in the proof of Proposition 7.5 is also a contraction in a metric ball of W_z of size δ, with δ small but fixed. It follows from uniqueness in the contraction mapping theorem that every critical point of $J_{K,\tau}$ in a δ-neighborhood of Z_τ must lie in $\widetilde{Z}_\tau$. By the blow-up analysis of the previous section, on $(S^3, g_{\mathbb{S}^3})$ every blowing-up solution is of the type given by Theorem 7.1. From a Taylor expansion on the reduced functional $F_\tau(\varphi_{q,\lambda})$, one can prove the uniqueness statement in Theorem 7.1.

(ii) In dimension $n \geq 4$, as we remarked above, it is possible to obtain solutions blowing up at multiple points. The strategy for their construction is similar, and uses a manifold Z_τ that depends on more parameters, describing the location, scales, and coefficients of each bubble function.

7.3 Blow-up analysis in arbitrary dimension

We conclude the chapter by stating a result on blow-up analysis in general dimension, but holding for special classes of diverging solutions. We will only give here some brief ideas of the proof, which relies on the above finite-dimensional reduction.

Theorem 7.11 ([165]). *Given (M^n, g) closed and of positive Yamabe class, suppose that $K\colon M \to \mathbb{R}$ is positive, Morse, and satisfies* (ND). *Let $\tau_i \to 0$, and $(u_i)_i$ be solutions of* $(\mathrm{E}_{K,\tau})$ *with $\tau = \tau_i$ such that $\|u_i\|_{H^1(M)} \leq C$ and $u_i \rightharpoonup 0$ as $i \to +\infty$. Then, if $\max_M u_i \to +\infty$, u_i only develops isolated simple blow-ups, which are in $\{\nabla K = 0\} \cap \{\Delta K < 0\}$.*

The main idea of the proof relies on the following two steps.

Step 1 (Struwe-type decomposition). If the solutions $(u_i)_i$ satisfy the above assumptions, there exists $L = L(C, K)$ such that u_i *splits* into q bubbles, with q uniformly upper bounded by a fixed constant $L > 0$, namely (for simplicity, we omit the index i in $p_{j,i}, \lambda_{j,i}$)

$$u_i = \sum_{j=1}^{q} \Big(\frac{K(p_j)}{n(n-1)} \Big)^{\frac{4}{n-2}} \varphi_{p_j,\lambda_j} + o_i(1) \quad \text{in } H^1(M),$$

with the property that

$$\frac{\lambda_j}{\lambda_\ell} + \frac{\lambda_\ell}{\lambda_j} + \lambda_j \lambda_\ell d^2(p_\ell, p_j) \to \infty \quad \text{as } i \to \infty, \tag{7.3}$$

similarly to Theorem 4.6; see Figure 7.4.

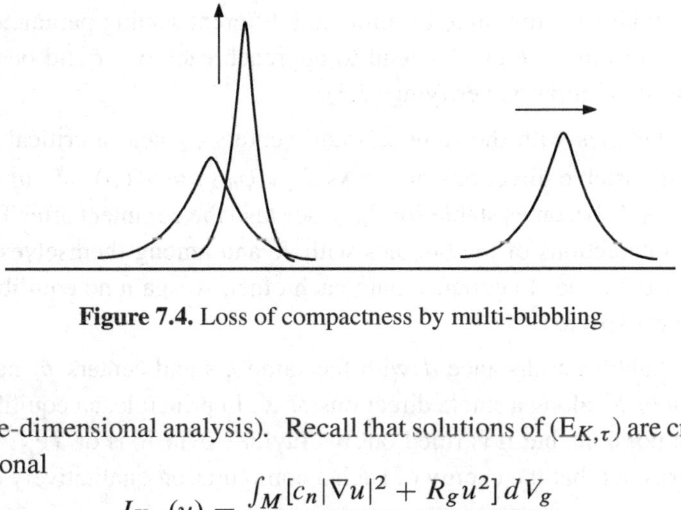

Figure 7.4. Loss of compactness by multi-bubbling

Step 2 (Finite-dimensional analysis). Recall that solutions of $(\mathrm{E}_{K,\tau})$ are critical points of the functional

$$J_{K,\tau}(u) = \frac{\int_M [c_n |\nabla u|^2 + R_g u^2] \, dV_g}{\big(\int_M K |u|^{2^*-\tau} \, dV_g\big)^{\frac{2}{2^*-\tau}}}.$$

Reasoning as for Remark 7.10, the solutions u_i are necessarily sums of the bubble functions in the above formula and a small correction w_τ as in Proposition 7.5. Neglecting, however, this correction w_τ and the coefficients in front of the φ_{p_j,λ_j}, we need naively to extremize $J_{K,\tau}(\sum_{j=1}^{q} \varphi_{p_j,\lambda_j})$ with respect to p_j and λ_j. Taylor-expanding the latter quantity, one observes *interaction terms* of the φ's with K, and among themselves, similarly to formulas (7.2) and (4.10). In particular, if two bubbles have the same scaling parameter λ and are at distance $\lambda^{-1} \ll d \ll 1$, there is a *potential energy* of order $-\lambda^{2-n} d^{2-n}$, which generates an *attractive force* between the two. The same is true for two bubble functions with the same center and different scaling parameters, which would tend to minimize the interaction energy for closer values of such parameters. The expansion for a single bubble at a critical point of K with negative Laplacian yields a function whose graph is pictured in Figure 7.3 (i).

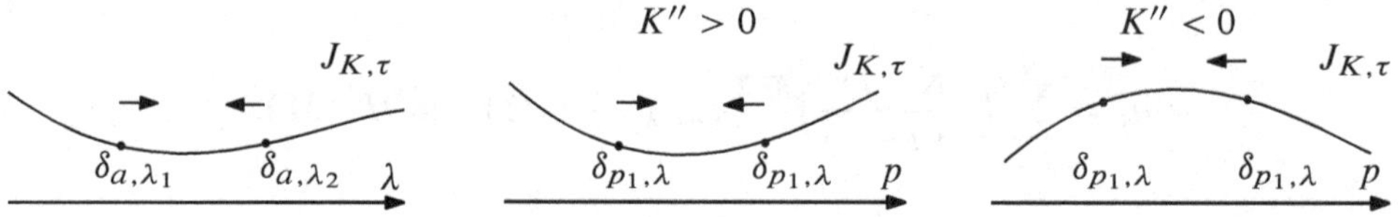

Figure 7.5. Interaction of two bubbles with different λ's or centers

We next qualitatively discuss three model cases, with two bubbles only (see Figure 7.5): we will argue that a multi-bubble configuration cannot be critical for $J_{K,\tau}$, and therefore blow-ups must be isolated simple.

(i) Two bubbles at the same location and different scaling parameters: $\varphi_{p,\lambda_1} + \varphi_{p,\lambda_2}$. In this case the λ's tend to approach each other and one expects no stable configuration verifying (7.3).

(ii) Two bubbles with the same λ's and centers p_i near a critical point of K, along unstable directions of K: As $J_{K,\tau}(\varphi_{p,\lambda}) \approx K(p)^{-\frac{2}{2^*}}$, unstable directions of K becomes stable for $J_{K,\tau}$; see also the comment after Theorem 7.1. The interactions of the bubbles with K and among themselves would lead the two bubbles to collapse onto each other, so again no equilibrium of this type is expected.

(iii) Two bubbles at distance d with the same λ's and centers p_i near a critical point of K, along a stable directions of K. In principle, an equilibrium is this time possible, but it is ruled out by Taylor expansions of $J_{K,\tau}$. In this case, it turns out that the energy of such a configuration qualitatively expands as

$$(1 - d^2)(\lambda^\tau + \lambda^{-2}) - (d\lambda)^{2-n}.$$

Extremizing in λ and in $d(p_1, p_2)$ one finds

$$\lambda^{-2} \simeq \tau + (\lambda d)^{2-n} \quad \text{and} \quad d \simeq \lambda^{2-n} d^{1-n},$$

a contradiction for λ large.

We finally derive the value of the limiting energy of a multi-bubble configuration. We will be sketchy in the proof, leaving the (simple) details to the reader.

Proposition 7.12. *Let (M, g) and K be as in Proposition 7.11. Let $\tau_i \searrow 0$ and let $(u_i)_i \rightharpoonup 0$ ($u_i = u_{\tau_i, x_1, \dots, x_q}$) be solutions of* $(\mathrm{E}_{K,\tau})$, $\tau = \tau_i$, *with isolated simple blow-ups at $x_1, \dots, x_q \in \{\nabla K = 0\} \cap \{\Delta K < 0\}$. Then*

$$J_{K,\tau_i}(u_i) \to \mathcal{Y}(\mathbb{S}^n, [g_{\mathbb{S}^n}])\Big(\sum_{j=1}^{q} K(x_j)^{\frac{2-n}{2}}\Big)^{\frac{2}{n}}.$$

Proof. We know that $u_i \approx \sum_{j=1}^{q} c_j \varphi_{x_j,\lambda_i}$, $\lambda_i \gg 1$, $\lambda_i^{\tau_i} \approx 1$ (for some suitable coefficients c_j). The function u_i solves $(\mathrm{E}_{K,\tau})$ with $\tau = \tau_i$ up to a Lagrange multiplier: $L_g u_i = \Lambda K u_i^{\frac{n+2}{n-2} - \tau_i}$. Without loss of generality, we can take $\Lambda = 1$, and assume that

$$L_g \varphi_{p,\lambda_i} = \varphi_{p,\lambda_i}^{\frac{n+2}{n-2}} = (1 + o_i(1))\varphi_{p,\lambda_i}^{\frac{n+2}{n-2} - \tau_i}.$$

Then, if $\gamma = \gamma_j > 0$, the function $\gamma \varphi_{p,\lambda_i}$ solves

$$L_g(\gamma \varphi_{p,\lambda_i}) = \gamma \varphi_{p,\lambda_i}^{\frac{n+2}{n-2}} = \gamma^{-\frac{4}{n-2}} (\gamma \varphi_{p,\lambda_i})^{\frac{n+2}{n-2}}.$$

To match this with $(\mathrm{E}_{K,\tau})$, we choose $p = q_j$ and $\gamma^{-\frac{4}{n-2}} = K(x_j)$, that is, $c_j = K(x_j)^{\frac{2-n}{4}}$. We then have

$$J_{K,\tau_i}(u_i) \approx \mathcal{Y}(\mathbb{S}^n, [g_{\mathbb{S}^n}]) \frac{\sum c_j^2}{(\sum c_j^{2^*} K(x_j))^{\frac{2}{2^*}}} = \mathcal{Y}(\mathbb{S}^n, [g_{\mathbb{S}^n}])\Big(\sum_{j=1}^{q} K(x_j)^{\frac{2-n}{2}}\Big)^{1-\frac{2}{2^*}},$$

and $1 - \frac{2}{2^*} = 1 - \frac{n-2}{n} = \frac{2}{n}$. This proves the desired assertion. ∎

Chapter 8

General existence results for (E_K)

In this chapter we will prove some existence results for (E_K) without requiring any symmetry assumptions, but under conditions involving the structure of critical points of the prescribed function K. We will crucially use the results of the previous two chapters, together with degree theory, min-max, and Morse-theoretical methods.

8.1 Proof of Theorem C

The main tool in the proof is *Leray–Schauder's degree*, for which we refer the reader to the books [10] and [81] for example.

The setting in which it can be employed is that of a Banach space X, an open bounded subset $\Omega \subseteq X$, and a continuous (non-linear) operator $S \in C(\bar{\Omega}; X)$ of the form $S = \mathrm{Id} - T$, where T is compact, meaning that T sends bounded sets into pre-compact ones. In such a case, for $p \notin S(\partial\Omega)$ (which is a closed set), one can define an integer $d_{\mathrm{LS}}(S, \Omega, p)$ that *counts with sign* the solutions x in Ω of $S(x) = p$. In finite dimensions, if such solutions are non-degenerate, the counting is locally given by the sign of the Jacobian determinant (see Figure 8.1 for the one-dimensional case), and, by an approximation procedure, it can be proven that the total number, the *Brouwer degree*, is well defined for general maps.

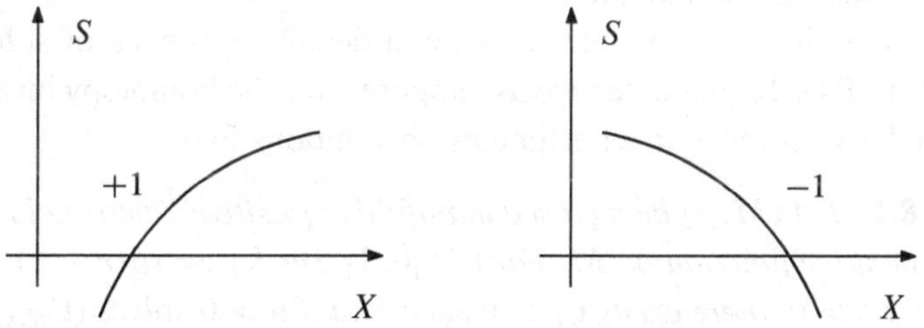

Figure 8.1. Local Leray–Schauder degree counting in one dimension

Its definition in infinite dimensions involves a further approximation process. By the compactness of T, one can take a family $(E_k)_k$ of finite-dimensional subspaces of X and a sequence of continuous maps $(T_k)_k$ with $T_k \to T$ uniformly in $\bar{\Omega}$, and $T_k(\bar{\Omega}) \subseteq E_k$. It can be shown that the limit of the Brouwer degrees of $\mathrm{Id} - T_k$ on $\Omega \cap E_k$ is well defined too, and called the *Leray–Schauder degree*, denoted by

$$d_{\mathrm{LS}}(S, \Omega, 0).$$

One can similarly define $d_{\mathrm{LS}}(S,\Omega,p)$ for $p \notin S(\partial\Omega)$ by counting with sign the solutions in Ω of $S(x) = p$. The following are the main properties of the Leray–Schauder degree:

(P1) (Solvability) If $d_{\mathrm{LS}}(S,\Omega,p) \neq 0$ then there exists $x \in \Omega$ such that $S(d) = p$.

(P2) (Homotopy invariance) If S_t, $t \in [0,1]$ is a continuous homotopy such that $S_t = \mathrm{Id} - T_t$ (T_t compact), $p \notin S_t(\partial\Omega)$ for every $t \in [0,1]$, then $d_{\mathrm{LS}}(S_t,\Omega,p)$ is independent of t.

(P3) (Excision) If $(\Omega_i)_i$, $i = 1,\dots,m$ are disjoint open sets of Ω such that $S(x) \neq p$ for every $x \in \bar{\Omega} \setminus \bigcup_{i=1}^m \Omega_i$, then

$$d_{\mathrm{LS}}(S,\Omega,p) = \sum_{i=1}^m d_{\mathrm{LS}}(S,\Omega_i,p).$$

(P4) (Local degree count) Suppose T is linear and compact, with $S = \mathrm{Id} - T$ invertible. Then for every $r > 0$ we have $d_{\mathrm{LS}}(S,B_r,0) = (-1)^{\sum m(\lambda_i)}$, where the λ_i are the (finitely many) negative eigenvalues of S, and $m(\lambda_i)$ stands for their algebraic multiplicity.

We will apply Leray–Schauder degree theory to study (E_K). The latter equation can fit into the above framework once we write it in the form

$$u - L_g^{-1}(K(x)u^{\frac{n+2}{n-2}}) = 0.$$

Working within positive functions in the Hölder space $C^{2,\alpha}(M)$, the operator on the left-hand side is of the form *identity – compact* by elliptic regularity theory (see Theorem 2.13) and Ascoli's theorem.

Later we will choose a suitable set Ω, and deduce existence of solutions using properties (P1)–(P4). To guarantee the assumptions for the homotopy invariance (P2), it is useful to derive some a priori estimates on solutions first.

Proposition 8.1. *Let (M,g) be a closed manifold of positive Yamabe class, and let K be a smooth positive function on M. For $t \in [0,1]$, set $K_t := tn(n-1) + (1-t)K$. Then, for every $\varepsilon > 0$, there exists $C_\varepsilon > 0$ such that if $u > 0$ solves ($E_{K,\tau}$) on M with $K = K_t$ for some $t \in [0,1]$ and $\tau \in [\varepsilon, \frac{4}{n-2} - \varepsilon]$, then*

$$u(x) \in (C_\varepsilon^{-1}, C_\varepsilon)\ \forall x \in M \quad \text{and} \quad \|u\|_{C^{2,\alpha}(M)} < C_\varepsilon.$$

Proof. We first prove the upper bound. Suppose by contradiction that there exist $\tau_i \geq \varepsilon$, $t_i \in [0,1]$ and solutions u_i of ($E_{K,\tau}$) with $K = K_{t_i}$, $\tau = \tau_i$, such that $\sup_M u_i \to \infty$. Let $x_i \in M$ be such that $M_i := u_i(x_i) = \sup_M u_i$. In geodesic coordinates at x_i, define the function v_i by

$$v_i(x) = M_i^{-1} u_i(M_i^{-\frac{p_i-1}{2}} x).$$

Then v_i satisfies

$$-c_n \Delta v_i = K_{t_i}(x_i) v_i^{\frac{n+2}{n-2}-\tau} + \text{l.o.t.} \quad \text{on } B_{R_i} \subseteq \mathbb{R}^n,\ R_i \to \infty.$$

Since $0 < v_i \le 1$, we see that $v_i \to \bar{v}$ in $C^2_{\text{loc}}(\mathbb{R}^n)$, where $\bar{v}$ solves

$$-c_n \Delta \bar{v} = K_{\bar{t}}(\bar{x}) \bar{v}^{\frac{n+2}{n-2}-\bar{\tau}} \quad \text{on } \mathbb{R}^n, \quad \bar{v}(0) = 1.$$

Here,

$$\bar{\tau} := \lim_{i\to\infty} \tau_i \ge \varepsilon, \quad \bar{x} := \lim_{i\to\infty} x_i, \quad \text{and} \quad \bar{t} := \lim_{i\to\infty} t_i \in [0,1].$$

Notice that $K_{\bar{t}}(\bar{x}) > 0$. By Theorem 5.8, $\bar{v} \equiv 0$, contradicting the fact that $\bar{v}(0) = 1$. This shows that $u_i \le C_\varepsilon$, giving the desired claim.

To prove $u_i \ge C_\varepsilon^{-1}$, we use Harnack's inequality (see e.g. [113, Theorem 8.8]) and the upper bound just shown. Finally, the bound $\|u\|_{C^{2,\alpha}(M)} < C_\varepsilon$ follows from Theorem 2.13. ∎

For (M, g) of positive Yamabe class, consider now the following situation: choose $X := C^{2,\alpha}(\mathbb{S}^3)$, $\tau = \varepsilon > 0$, C_ε as in Proposition 8.1, and let

$$\Omega := \{u \in X : u(x) \in (C_\varepsilon^{-1}, C_\varepsilon)\ \forall x \in M,\ \|u\|_{C^{2,\alpha}(M)} < C_\varepsilon\}.$$

By definition, Ω is open and bounded in X. For $t \in [0,1]$, consider the operators $S_t : X \to X$ defined by

$$S_t u = u - L_g^{-1}(K_t |u|^{\frac{n+2}{n-2}-\tau}). \tag{8.1}$$

As remarked before, S_t is of the form *identity – compact*.

Corollary 8.2. *If Ω and S_t are as above, the Leray–Schauder degree $d_{\text{LS}}(S_t, \Omega, 0)$ is well defined and independent of $t \in [0,1]$.*

Proof. Proposition 8.1 implies that $0 \notin S_t(\partial\Omega)$ for every $t \in [0,1]$. The corollary follows then from property (P2) of the Leray–Schauder degree. ∎

Lemma 8.3. *Let $J_{K,\tau}$ be as in (6.1), and suppose v is a non-degenerate (up to dilations) critical point of $J_{K,\tau}$, with Morse index m. Let $\mu > 0$ be such that $\tilde{v} = \mu v$ solves $(\mathrm{E}_{K,\tau})$. Then there exists $r > 0$ small such that, letting S_t be as in (8.1), one has $d_{\text{LS}}(S_0, B_r(\tilde{v}), 0) = (-1)^{m+1}$.*

Proof. Consider the sphere $\Sigma_{\tilde{v}} = \{u \in H^1(\mathbb{S}^3) : \|u\| = \|\tilde{v}\|\}$. Then $\tilde{v}$ is non-degenerate for $J_{K,\tau}|_{\Sigma_{\tilde{v}}}$, with m negative directions in $T_{\tilde{v}}\Sigma_{\tilde{v}}$ for its Hessian. On the other hand, the non-homogeneous functional

$$I_{K,\tau}(u) := \frac{1}{2}\int_M [c_n |\nabla u|^2 + R_g u^2]\, dV_g - \frac{1}{2^* - \tau}\int_M K|u|^{2^*-\tau}$$

has one more negative direction (parallel to $\tilde{v}$); see Figure 8.2. ∎

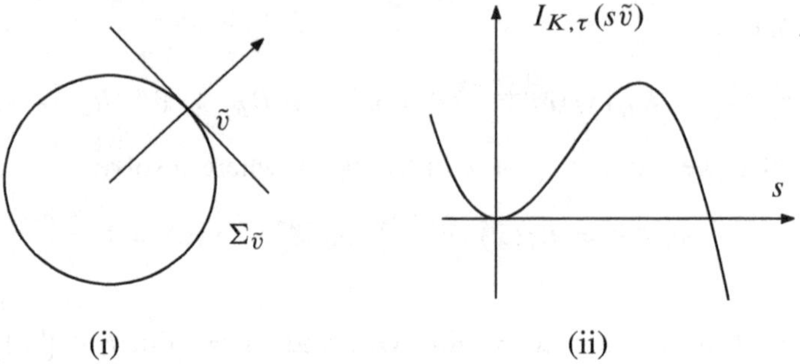

Figure 8.2. A spherical constraint and the radial behavior of $I_{K,\tau}$

This yields the relation $\mathrm{ind}(I_{K,\tau}, \tilde{v}) = m + 1$, which implies that the linearization of (E$_{K,\tau}$) at $\tilde{v}$ has $m + 1$ negative eigenvalues. By property (P4) of d_{LS} we have that the linearization of S_0 at $\tilde{v}$ satisfies

$$d_{\mathrm{LS}}(dS_0(\tilde{v}), B_r(0), 0) = (-1)^{m+1} \quad \text{for every } r > 0.$$

For $r > 0$ small, using (P2) and a Taylor expansion, namely

$$S_0(u) = \underbrace{S_0(\tilde{v})}_{=0} + dS_0(\tilde{v})[u - \tilde{v}] + O(\|u - \tilde{v}\|^2),$$

we find also that $d_{\mathrm{LS}}(S_0, B_r(\tilde{v}), 0) = (-1)^{m+1}$, as desired.

The next rigidity result is useful to compute the Leray–Schauder degree of S_t in Ω starting from $t = 1$.

Proposition 8.4 ([34, 112, 181]). *Suppose (M^n, g) is a closed Einstein manifold of positive Yamabe class, and let $u > 0$ solve* (E$_{K,\tau}$) *with $K \equiv 1$ and $\tau > 0$. Then $u \equiv$ constant. If (M^n, g) is not conformally equivalent to $(\mathbb{S}^n, g_{\mathbb{S}^n})$, the result is also true for $\tau = 0$.*

A more general version of the above proposition is Theorem A.5, stated and proved in the appendix.

Corollary 8.5. *For $(M, g) = (\mathbb{S}^n, g_{\mathbb{S}^n})$, if Ω is as above, then $d_{\mathrm{LS}}(S_0, \Omega, 0) = -1$.*

Proof. The round sphere has constant scalar curvature identically equal to $n(n-1)$. By Proposition 8.4, the only solution of (E$_{K,\tau}$) with $K \equiv 1$ is

$$u \equiv (n(n-1))^{\frac{1}{\frac{4}{n-2}-\tau}} \in \Omega.$$

The linearization of (E$_{K,\tau}$) at $u \equiv \bar{R}^{\frac{4}{n-2}-\tau}$ is

$$-4\frac{n-1}{n-2}\Delta v + n(n-1)v - \Big(\frac{n+2}{n-2} - \tau\Big)n(n-1)v,$$

that is,

$$4\frac{n-1}{n-2}(-\Delta v + nv) + n(n-1)\tau v.$$

On $\mathbb{S}^n$ and for $\tau > 0$ small, this operator possesses only one negative eigenvalue and trivial kernel; see [9, Chapter 4]. Hence by property (P3) we find that for $r > 0$ sufficiently small,

$$d_{\mathrm{LS}}(S_0, \Omega, 0) = d_{\mathrm{LS}}\Big(S_0, B_r\Big(u \equiv (n(n-1))^{\frac{1}{\frac{4}{n-2}-\tau}}\Big), 0\Big) = -1,$$

concluding the proof. ■

In Section 6.2 and Chapter 7 we saw that blowing-up solutions of $(\mathrm{E}_{K,\tau})$ are in one-to-one correspondence with critical points of K such that $\Delta K < 0$, and this allows one to compute their contribution to the degree.

Proof of Theorem C. We proved that on $\mathbb{S}^3$, all blow-ups of solutions to $(\mathrm{E}_{K,\tau})$ as $\tau \to 0$ are isolated simple. Therefore, for any $\delta > 0$ there exists $D_\delta > 0$ such that if u solves $(\mathrm{E}_{K,\tau})$ with $\tau \leq \delta$, then either $u(x) \in (D_\delta^{-1}, D_\delta)$ for every $x \in \mathbb{S}^3$ and $\|u\|_{C^{2,\alpha}(\mathbb{S}^3)} < D_\delta$, or u develops an isolated simple blow-up.

Letting

$$\tilde{\Omega} := \big\{u \in X : u(x) \in (D_\delta^{-1}, D_\delta)\ \forall x \in \mathbb{S}^3,\ \|u\|_{C^{2,\alpha}(\mathbb{S}^3)} < D_\delta\big\},$$

we have the dichotomy for solutions of $(\mathrm{E}_{K,\tau})$, $\tau \in (0, \delta)$, as pictured in Figure 8.3. This implies that $d_{\mathrm{LS}}(S, \tilde{\Omega}, 0)$ is independent of $\tau \in (0, \delta)$. We also know from the

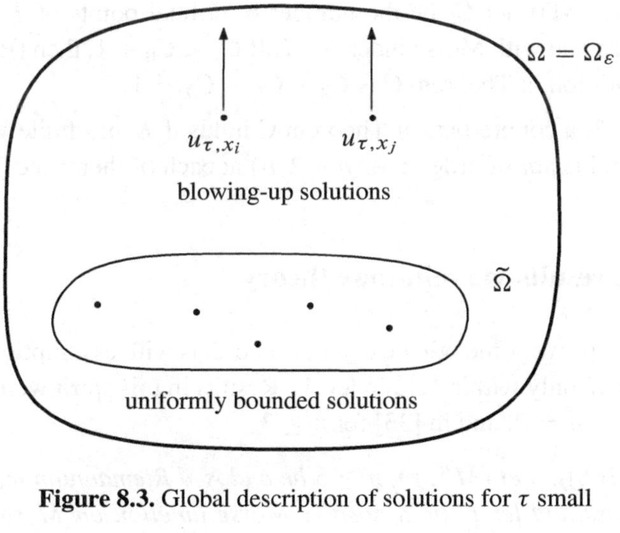

Figure 8.3. Global description of solutions for τ small

results in Chapter 7 that for every $\bar{x} \in \mathbb{S}^3$ with $\nabla K(\bar{x}) = 0$ and $\Delta K(\bar{x}) < 0$ there exists a unique solution $u_{\tau,\bar{x}}$ blowing up at $\bar{x}$, and moreover $\operatorname{ind}(J_{K,\tau}, u_{\tau,\bar{x}}) = 3 - \operatorname{ind}(K, \bar{x})$.

By Lemma 8.3, there exists $r > 0$ small such that

$$d_{\mathrm{LS}}(S = S_1, B_r(u_{\tau,\bar{x}}), 0) = (-1)^{1+3-\operatorname{ind}(K,\bar{x})} = (-1)^{\operatorname{ind}(K,\bar{x})}.$$

For $\tau < \varepsilon \ll \delta$, let C_ε be as in Proposition 8.1, and let us again consider the set

$$\Omega := \{u \in X : u(x) \in (C_\varepsilon^{-1}, C_\varepsilon)\ \forall x \in \mathbb{S}^3,\ \|u\|_{C^{2,\alpha}(\mathbb{S}^3)} < C_\varepsilon\},$$

containing both $\widetilde{\Omega}$ and the u_{τ,x_i}. Then for $\tau \in (0, \delta)$, $d_{\mathrm{LS}}(S, \widetilde{\Omega}, 0)$ (independent of τ) equals $d_{\mathrm{LS}}(S, \widetilde{\Omega}, 0)_{\tau=\varepsilon}$. By the excision property (P3), we have

$$d_{\mathrm{LS}}(S, \widetilde{\Omega}, 0)|_{\tau<\varepsilon} = d_{\mathrm{LS}}(S, \Omega, 0)|_{\tau<\varepsilon} - \sum_{\substack{\nabla K(x_i)=0 \\ \Delta K(x_i)<0}} (-1)^{\operatorname{ind}(K,x_i)}.$$

By Corollaries 8.2 and 8.5,

$$d_{\mathrm{LS}}(S, \widetilde{\Omega}, 0)|_{\tau<\varepsilon} = -1 - \sum_{\substack{\nabla K(x_i)=0 \\ \Delta K(x_i)<0}} (-1)^{\operatorname{ind}(K,x_i)} \neq 0.$$

By the solvability property (P1), for every $\tau \in (0, \delta)$ there exists a solution of ($E_{K,\tau}$) in $\widetilde{\Omega}$. Passing to the limit $\tau \to 0$, these converge to a positive solution of (E_K). ∎

Remark 8.6. We have the following more general results:

(i) Theorem C was extended in [202] using Morse theory; see also [163]. Still assuming (ND), let C_i be the number of critical points of K with negative Laplacian and with Morse index $3 - i$. If $C_1 < C_0 + 1$, then (E_K) is solvable. The condition in Theorem C is $C_2 - C_1 + C_0 \neq 1$.

(ii) For $n > 3$, a counterpart of Theorem C holds if K has finitely many critical points and is *flat* of order $\beta \in (n-2, n)$ at each of them; see [158].

8.2 Existence results via min-max theory

In this section we prove some other existence theorems with assumptions that involve critical points of K only within certain levels. Results in this spirit were first obtained in [60, 66, 208] for $n = 2$, and in [33] for $n \geq 3$.

Theorem 8.7 ([166]). *Let (M^n, g), $n \geq 5$ be a closed Riemannian manifold of positive Yamabe class and let K be a positive Morse function on M satisfying* (ND).

Suppose K has a local maximum point z, and that there exists a curve $a(t)$ joining z to another point y with

$$K(y) \geq K(z),$$

such that both the following properties hold:

(i) *for all $x_i \neq x_j$ local maxima of K,*

$$\max_t K(a(t))^{\frac{2-n}{n}} < \left(K(x_i)^{\frac{2-n}{2}} + K(x_j)^{\frac{2-n}{2}}\right)^{\frac{2}{n}};$$

(i) *critical points of index $n-1$ for which K is in the range*

$$\left[\min_t K(a(t)), K(z)\right]$$

have positive Laplacian.

Then (E_K) *admits a positive solution.*

Proof. One can easily see that there exists a curve $\tilde{a}(t)$ joining y to another maximum point $\tilde{z}$ of K and such that $\min_t K(\tilde{a}(t)) = K(y)$. By Theorem 7.1, there exist solutions $u_{\tau,z}$, $u_{\tau,\tilde{z}}$ of (E_K) that are local minima of $J_{K,\tau}$.

Next define the composition $\hat{a} := a * \tilde{a}$, which is a path joining z to $\tilde{z}$. It is possible to construct a one-parameter family of functions $u(t)$ close in $H^1(M)$ to $\varphi_{\hat{a}(t),\lambda(t)}$ with $\lambda(t) \simeq \tau^{-\frac{1}{2}}$ and $u(0) = u_{\tau,z}$, $u(1) = u_{\tau,\tilde{z}}$. It is easy to show that, for τ small enough, $J_{K,\tau}(u(t)) = Y(\mathbb{S}^n, [g_{\mathbb{S}^n}]) K(\hat{a}(t))^{\frac{2-n}{n}} + o_\tau(1)$, and therefore

$$\max_{t\in[0,1]} J_{K,\tau}(u(t)) \leq Y(\mathbb{S}^n, [g_{\mathbb{S}^n}])\big(\min_{t\in[0,1]} K(a(t))\big)^{\frac{2-n}{n}} + o_\tau(1).$$

Consider the mountain-pass level between the (strict) local minima $u_{\tau,z}$, $u_{\tau,\tilde{z}}$ of J_τ. Assuming that (E_K) has no solutions, by the Palais–Smale condition for J_τ and by the fact that all critical points with uniformly bounded energy of J_τ as described in Theorem 7.11 are non-degenerate (see also Remark 7.2 (ii)), J_τ must possess a critical point *of index* 1 at a level less than or equal to

$$Y(\mathbb{S}^n, [g_{\mathbb{S}^n}])\big(\min_{t\in[0,1]} K(a(t))\big)^{\frac{2-n}{n}} + o_\tau(1).$$

By the same remark, we must either have a simple blow-up at a critical point p of K of index $n-1$ with

$$K(p) \in \left[\min_t K(a(t)), K(z)\right],$$

or a double blow-up at two local maxima x_1, x_2 of K such that

$$\max_t K(a(t))^{\frac{2-n}{n}} \geq \left(K(x_1)^{\frac{2-n}{2}} + K(x_1)^{\frac{2-n}{2}}\right)^{\frac{2}{n}};$$

see Proposition 7.5. However by our assumptions none of these possibilities is allowed, and we reach a contradiction. ■

Remark 8.8. We notice the following facts:

(i) The latter result improves the pinching condition of [33, Theorem 2] (if compactified from $\mathbb{R}^n$ to $\mathbb{S}^n$) for K Morse and satisfying (ND), namely

  (j) $K_{\max} < 2^{\frac{2}{n-2}} \min_t K(x(t))$;

  (jj) critical points in the range $[\min_t K(x(t)), K(z))$ are local maxima or have positive Laplacian.

(ii) On $(\mathbb{S}^3, g_{\mathbb{S}^3})$, Theorem 8.7 holds true without condition (i). This follows from the fact that on the round three-sphere, only single blow-ups may occur; see Theorem 6.9. In four dimensions one could relax condition (i) in Theorem 8.7, taking Remark 6.16 (i) into account.

Next we state a related result, proved with similar techniques.

Theorem 8.9 ([166]). *Let (M^n, g), $n \geq 5$ be a closed Riemannian manifold of positive Yamabe class and let K be a positive Morse function on M satisfying* (ND). *Assume that there is a set $\Xi \subseteq M$ with $\mathfrak{C}$ components that contains p local maxima $x_1, \dots, x_p$ of K and such that*

$$\max_{\Xi}(K^{\frac{2-n}{2}}) < \min\big\{(K(x_i)^{\frac{2-n}{2}} + K(x_j)^{\frac{2-n}{2}})^{\frac{2}{n}} : x_i \neq x_j \text{ local maxima of } K\big\}.$$

Assume also that K has $q \geq 0$ critical points of index 1 in the range

$$\big[\min_{\Xi} K, \max_{i\in\{1,\dots,p\}} K(x_i)\big).$$

Then (E$_K$) *has a solution provided that $q < p - \mathfrak{C}$.*

Proof. Arguing by contradiction and assuming that (E$_K$) is not solvable, all solutions of (E$_{K,\tau}$) with uniformly bounded energy must have zero weak limit. Fix $\varepsilon > 0$ small; we know by Theorem 7.1 that J_τ has at least p local minima $u_{\tau,x_1}, \dots, u_{\tau,x_p}$ such that for τ small there holds

$$J_\tau(u_{\tau,x_j}) < Y(\mathbb{S}^n, [g_{\mathbb{S}^n}])\big(\min_{i\in\{1,\dots,p\}} K(x_i)\big)^{\frac{2-n}{n}} + \varepsilon$$

and such that, for all sufficiently small values of τ, J_τ has no critical point at level

$$Y(\mathbb{S}^n, [g_{\mathbb{S}^n}])\big(\min_{i\in\{1,\dots,p\}} K(x_i)\big)^{\frac{2-n}{n}} + \varepsilon.$$

We can assume that for τ small there is no critical point of J_τ at level

$$Y(\mathbb{S}^n, [g_{\mathbb{S}^n}]) \max_{\Xi}(K^{\frac{2-n}{2}}) + \varepsilon,$$

and we can modify J_τ near all its local minima at level less than or equal to

$$Y(\mathbb{S}^n, [g_{\mathbb{S}^n}]) \max_{\Xi}(K^{\frac{2-n}{2}}) + \varepsilon,$$

which are non-degenerate (see Remark 6.16), in order to still have the Palais–Smale condition, to not generate new critical points, and so that the modified minima are at level zero. Call $\tilde{J}_\tau$ the resulting functional, which we can take of class $C^{2,\alpha}$ as the original one; see Figure 8.4. The functional $\tilde{J}_\tau$ will also possess at least p critical points at level zero.

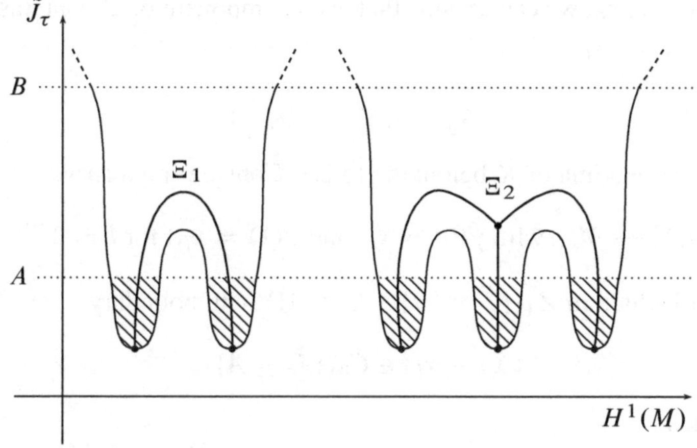

Figure 8.4. The modified functional $\tilde{J}_\tau$ and its sub-levels

We then use *relative Morse inequalities* for $\tilde{J}_\tau$ [53, Theorem 4.3] between the levels

$$A = \varepsilon \quad \text{and} \quad B = Y(\mathbb{S}^n, [g_{\mathbb{S}^n}]) \max_{\Xi}(K^{\frac{2-n}{2}}) + \varepsilon.$$

By construction, in the range $[A, B]$, $\tilde{J}_\tau$ has $C_0 = 0$ critical points of index 0 and $C_1 = q$ critical points of index 1. Since $\tilde{J}_\tau$ has no local minima in the range $[A, B]$ and the Palais–Smale condition holds true, every point of $\{\tilde{J}_\tau \leq B\}$ can be joined to $\{\tilde{J}_\tau \leq A\}$ within $\{\tilde{J}_\tau \leq B\}$. As a consequence, considering relative homologies one has

$$\beta_0 := \operatorname{rank} H_0\big(\{\tilde{J}_\tau \leq B\}, \{\tilde{J}_\tau \leq A\}\big) = 0;$$

see e.g. [127, Chapter 2, Exercise 16, p. 130]. On the other hand, consider

$$\beta_1 := \operatorname{rank} H_1\big(\{\tilde{J}_\tau \leq B\}, \{\tilde{J}_\tau \leq A\}\big).$$

Recall that

$$H_1(\{\tilde{J}_\tau \le B\}, \{\tilde{J}_\tau \le A\}) = Z_1(\{\tilde{J}_\tau \le B\}, \{\tilde{J}_\tau \le A\})/B_1(\{\tilde{J}_\tau \le B\}, \{\tilde{J}_\tau \le A\}),$$

where Z_1 and B_1 denote kernel and image of the boundary operators in one and two homological dimensions respectively; see [169, Chapter VII, §6]. We claim next that

$$\beta_1 \ge p - \mathfrak{C}. \tag{8.2}$$

To prove this, let $\Xi_1, \dots, \Xi_{\mathfrak{C}}$ denote the connected components of Ξ. As our assumptions improve or stay invariant if we remove components containing no or only one point from $x_1, \dots, x_p$, we can assume that each component of Ξ contains at least two points from $x_1, \dots, x_p$.

Given Ξ_j, let

$$X_j = \{x_{i_1}, \dots, x_{i_{\mathfrak{C}_j}}\}$$

denote the local maxima of K belonging to Ξ_j. Considering a curve

$$\gamma_{j,l}\colon [0,1] \to M \quad \text{with } \gamma(0) = x_{i_1} \text{ and } \gamma(1) = x_{i_l}, \text{ for } l = 2, \dots, i_{\mathfrak{C}_j},$$

its image is a 1-chain in $Z_1(\{\tilde{J}_\tau \le B\}, \{\tilde{J}_\tau \le A\})$ with boundary

$$x_{i_l} - x_{i_1} \in C_0(\{\tilde{J}_\tau \le A\}).$$

It turns out that

$$\gamma_{j,2}, \dots, \gamma_{j,\mathfrak{C}_j} \text{ generate } \mathfrak{C}_j - 1 \text{ elements of } H_1(\{\tilde{J}_\tau \le B\}, \{\tilde{J}_\tau \le A\}), \tag{8.3}$$

which are linearly independent. To prove (8.2) we show that any linear combination

$$\sum_{2 \le h \le \mathfrak{C}_j} n_h \gamma_{j,h}$$

with not all $n_h = 0$ cannot be written as

$$\sum_{2 \le h \le \mathfrak{C}_j} n_h \gamma_{j,h} = \partial_2 \mathrm{c}_2 + \mathrm{c}_1 \tag{8.4}$$

with $\mathrm{c}_2 \in C_2(\{\tilde{J}_\tau \le B\}, \{\tilde{J}_\tau \le A\})$ and $\mathrm{c}_1 \in C_1(\{\tilde{J}_\tau \le A\})$.

In fact, let us apply the boundary operator ∂_1 to both sides of the latter equation. Since not all n_h are zero, $\partial_1(\sum_h n_h \gamma_{j,h})$ is non-trivial in $C_0(\{\tilde{J}_\tau \le A\})$. Clearly $\partial_1 \circ \partial_2 = 0$, so to achieve (8.4) we would need $\partial_1 \mathrm{c}_1$ to be in $C_0(\{\tilde{J}_\tau \le A\})$, a non-trivial linear combination of the points $x_{i_1}, \dots, x_{i_l}$. However, since $x_{i_1}, \dots, x_{i_l}$ lie in different components of $\{\tilde{J}_\tau \le A\}$, there is no chain $\mathrm{c}_1 \in C_1(\{\tilde{J}_\tau \le A\})$ with this

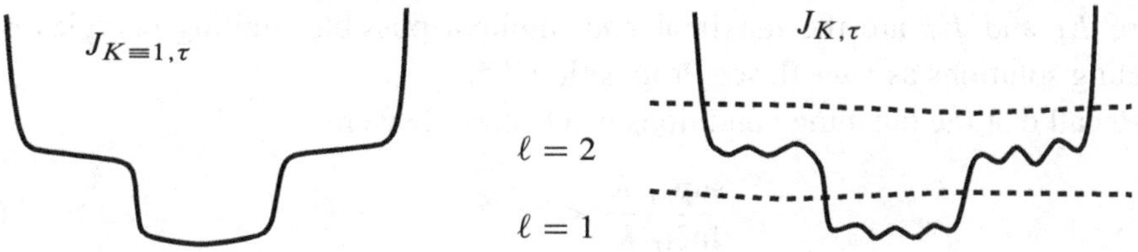

Figure 8.5. Geometry of $J_{K,\tau}$ for K pinched

property. This shows (8.3). Repeating this reasoning for every component of Ξ, we obtain

$$\beta_1 \geq \sum_{j=1}^{\mathfrak{C}} (\mathfrak{C}_j - 1) = p - \mathfrak{C},$$

since $\sum_{j=1}^{\mathfrak{C}} \mathfrak{C}_j = p$. This shows (8.2) as well. Now, the relative Morse inequalities imply

$$q = C_1 = C_1 - C_0 \geq \beta_1 - \beta_0 \geq p - \mathfrak{C},$$

contradicting our assumptions. ∎

Corollary 8.10. *Suppose that $K\colon \mathbb{S}^n \to \mathbb{R}_+$ satisfies $\frac{K_{\max}}{K_{\min}} \leq 2^{\frac{2}{n-2}}$ and that it has p local maxima and q critical points of index $n-1$ with negative Laplacian. Then* (E_K) *admits a positive solution provided $q < p - 1$.*

Proof. In the previous theorem it is sufficient to choose the connected set $\Xi = \mathbb{S}^n$. ∎

8.3 Proof of Theorem D

For the proof we will exploit the fact that for K close to a constant, say $K \approx 1$, the limit energy of a solution forming ℓ bubbles is nearly proportional to $\ell^{\frac{2}{n}}$, see Proposition 7.5, and therefore the energy picture will become as in Figure 8.5. We will be able to compute the contribution to the total index of solutions via the *Poincaré–Hopf theorem.*

Up to renaming the indices, we can label the critical points of K with negative Laplacian (i.e., the possible blow-up points) as $\{x_1, x_2, \dots, x_q\}$ so that

$$K(x_1) \geq K(x_2) \geq \cdots \geq K(x_q).$$

Then, for $1 \leq \ell \leq q$, we set

$$\underline{E}_\ell := \mathcal{Y}(\mathbb{S}^n, [g_{\mathbb{S}^n}])\Big(\sum_{i=1}^{\ell} K(x_i)^{\frac{2-n}{2}}\Big)^{\frac{2}{n}}, \quad \bar{E}_\ell := \mathcal{Y}(\mathbb{S}^n, [g_{\mathbb{S}^n}])\Big(\sum_{i=q-\ell+1}^{q} K(x_i)^{\frac{2-n}{2}}\Big)^{\frac{2}{n}},$$

where $\bar{E}_\ell$ and $\underline{E}_\ell$ are the maximal and minimal possible limiting energies of ℓ-bubbling solutions as $\tau \to 0$; see Proposition 7.5.

Recall that the pinching conditions in Theorem D were

$$\frac{\sup_M K}{\inf_M K} < 2^{\frac{1}{n-2}}, \tag{P1}$$

$$\frac{\sup_M K}{\inf_M K} < \left(\frac{3}{2}\right)^{\frac{1}{n-2}}. \tag{P2}$$

Proposition 8.11. *Let* (M, g) *and* K *be as in Theorem* D. *For* ℓ *integer, let* $\bar{E}_{\ell,\tau}$ *and* $\underline{E}_{\ell,\tau}$ *be as above. Assume that* (E_K) *has no solution, namely that all solutions of* ($E_{K,\tau}$) *blow up as* $\tau \to 0$. *If* (P1) *holds, for* τ *small there exists* $\alpha \in (\bar{E}_{1,\tau}, \underline{E}_{2,\tau})$ *such that* $\{J_{K,\tau} < \alpha\}$ *is contractible. If* (P2) *holds, we have a similar conclusion for some* $\beta \in (\bar{E}_{2,\tau}, \underline{E}_{3,\tau})$.

Proof. For ℓ integer, let $\bar{E}_{\ell,\tau}$ be the maximal energy of an ℓ-bubble blowing-up solution and $\underline{E}_{\ell,\tau}$ be the minimal one.

For convenience we write J_τ and $\bar{J}_\tau$ for $J_{K,\tau}$ and $J_{K\equiv 1,\tau}$ respectively. We claim that if

$$\{\partial J_\tau = 0\} \cap \{\underline{A} \le J_\tau \le \bar{A}\} = \emptyset, \quad \bar{A} := \left(\frac{K_{\max}}{K_{\min}}\right)^{\frac{2}{p+1}}, \; p := \frac{n+2}{n-2} - \tau,$$

for some $\underline{A} > 0$ then the sub-level $\{J_\tau \le c\}$ is contractible for every $c \in [\underline{A}, \bar{A}]$. Once the claim is proven, the proposition will follow.

By a direct computation one finds

$$J_\tau(u) \ge K_{\max}^{-\frac{2}{p+1}} \bar{J}_\tau(u) \quad \text{and} \quad J_\tau(u) \le K_{\min}^{-\frac{2}{p+1}} \bar{J}_\tau(u),$$

and hence

$$\{J_\tau \le \underline{A}\} \subseteq \{\bar{J}_\tau \le K_{\max}^{\frac{2}{p+1}} \underline{A}\} \subseteq \{J_\tau \le \bar{A}\}.$$

As the norm on dJ_τ is uniformly bounded on a given sub-level of J_τ, and since the second differential is locally Hölder continuous, the negative gradient flow $\dot{\phi} = -\nabla J_\tau$ is globally well defined, and depends continuously on the initial condition. Since J_τ satisfies the Palais–Smale condition for each $\tau > 0$, and

$$\{\partial J_\tau = 0\} \cap \{\underline{A} \le J_\tau \le \bar{A}\} = \emptyset,$$

by the deformation lemma (see [10] or [209]) for any $u \in \{\underline{A} \le J_\tau \le \bar{A}\}$, there exists a first time $T_u \ge 0$ (depending continuously on u) such that $\phi(u, T_u) \in \{J_\tau \le \underline{A}\}$. We now consider the homotopy

$$F(s, u) := \phi(sT_u, u).$$

If $u \in \{J_\tau \leq \underline{A}\}$ then we see that $T_u = 0$, and therefore, $F(s,u) = u$ for every $s \in [0,1]$. This shows that F deforms $\{\bar{J}_\tau \leq K_{\max}^{2/(p+1)}\}$ into $\{J_\tau \leq \underline{A}\}$, but not necessarily within $\{\bar{J}_\tau \leq K_{\max}^{2/(p+1)}\}$. However, this can be achieved by composing ϕ on the left by a suitable Yamabe-type flow, namely the gradient flow of $\bar{J}_\tau$.

Recall from Proposition 8.4 that if $\tau > 0$ and (M,g) is Einstein (so that R_g is constant), then $(\mathrm{E}_{K,\tau})$ with $K \equiv 1$ has only constant solutions. Hence the infimum of $\bar{J}_\tau$ is attained and equal to $R_g \mathrm{Vol}_g(M)^{1-2/(p+1)}$. Since $\bar{J}_\tau$ also satisfies the Palais–Smale condition for every $\tau > 0$, the negative gradient flow $\bar{\phi}(t,u)$ of $\bar{J}_\tau$ evolves all initial data u to a constant function, intersecting transversally every level set of $\bar{J}_\tau$ higher than its infimum. As before, for every initial datum u there exists a first time $\bar{T}_u \geq 0$ such that

$$\bar{\phi}(\bar{T}_u, u) \in \{\bar{J}_\tau \leq K_{\max}^{\frac{2}{p+1}} \underline{A}\}.$$

Defining

$$\widetilde{F}(s,u) := \bar{\phi}(\bar{T}_{F(s,u)}, F(s,u)),$$

we deduce that $\widetilde{F}$ is a deformation retract of $\{\bar{J}_\tau \leq K_{\max}^{2/(p+1)} \underline{A}\}$ onto $\{J_\tau \leq \underline{A}\}$, and therefore it realizes a homotopy equivalence. Since the deformation lemma and the Palais–Smale condition for $\bar{J}_\tau$ imply that $\{\bar{J}_\tau \leq K_{\max}^{2/(p+1)} \underline{A}\}$ is homotopically equivalent to a point, the same property holds for $\{J_\tau \leq \underline{A}\}$. In fact, the same conclusion holds for $\{J_\tau \leq c\}$ if $c \in [\underline{A}, \bar{A}]$.

The proof is similar replacing (P1) with (P2) and α with β. ■

We will then need the following well-known result (see e.g. [53]) relating the structure of critical points of a Morse function to the Euler characteristic of its domain.

Theorem 8.12 (Poincaré–Hopf). *Suppose $X \subseteq H$ is a Hilbert sub-manifold of a Hilbert space H, and let $f: X \to \mathbb{R}$ be of class C^2, bounded from below, with second differential of the form $d^2 f(x) = \mathrm{Id} -$ compact on $T_x X$ for every $x \in X$, and assume that f satisfies the Palais–Smale condition. Suppose f has no critical point at level a and that f is Morse with only finitely many critical points $(z_i)_i$. Then, letting $\Omega = \{f < a\}$ and assuming it is bounded, one has*

$$\sum_{z_i \in \{\nabla f = 0\} \cap \{f < a\}} (-1)^{\mathrm{ind}(f, z_i)} = \chi(\Omega).$$

We apply this result to $X = \{\|u\|_{H^1(M)} = 1\}$ and $f = J_{K,\tau}$, arguing by contradiction and assuming that (E_K) has no solutions. To be more precise, in order to remove the degeneracy of $J_{K,\tau}$ by scaling (recall that it is homogeneous of order zero) one could add to it a function of $\|u\|_{H^1(M)}$ possessing only one critical point of minimum type and which is arbitrarily small. Without writing this explicitly, we assume this observation will be understood.

To prove the theorem under the hypotheses (P1) and ($*$), we use the Poincaré–Hopf theorem with $a = \alpha$ and $\chi(\Omega) = 1$. Then

$$1 = \chi(\Omega) = \operatorname{ind}(\nabla J_{K,\tau}, \Omega, 0) = \sum_{x_i \in \{\nabla K = 0\} \cap \{\Delta K < 0\}} (-1)^{n - \operatorname{ind}(K, x_i)} \neq 1,$$

a contradiction. Hence, $J_{K,\tau}$ has other critical points in Ω, that must have a non-trivial limit as $\tau \to 0$.

We now continue the proof under hypothesis (P2) and the assumption that there are at least two critical points of K with negative Laplacian. Notice that (P2) implies (P1), so we can assume that ($*$) does not hold, that is

$$\sum_{x_i \in \{\nabla K = 0\} \cap \{\Delta K < 0\}} (-1)^{n - \operatorname{ind}(K, x_i)} = (-1)^n.$$

Otherwise we are in the previous case.

The index contribution of 2-bubbles must vanish, that is (the sum is on the finite set $\{\nabla K = 0\} \cap \{\Delta K < 0\}$)

$$0 = \sum_{i<j} (-1)^{\operatorname{ind}(J_{K,\tau}, u_{\tau, x_i, x_j})}.$$

Recall that, by Remark 7.2 (ii),

$$\begin{aligned} \operatorname{ind}(J_{K,\tau}, u_{\tau, x_i, x_j}) &= 1 + (n - \operatorname{ind}(K, x_i)) + (n - \operatorname{ind}(K, x_j)) \\ &= 2n + 1 - \operatorname{ind}(K, x_i) - \operatorname{ind}(K, x_j), \end{aligned}$$

and therefore,

$$(-1)^{\operatorname{ind}(J_{K,\tau}, u_{\tau, x_i, x_j})} = (-1)^{\operatorname{ind}(K, x_i) + \operatorname{ind}(K, x_j) + 1}.$$

Hence we obtain

$$\begin{aligned} 0 &= -\sum_{i<j} (-1)^{\operatorname{ind}(J_{K,\tau}, u_{\tau, x_i, x_j})} = \frac{1}{2} \sum_{i \neq j} (-1)^{\operatorname{ind}(K, x_i) + \operatorname{ind}(K, x_j)} \\ &= \frac{1}{2} \sum_i \Big(\sum_{j \neq i} (-1)^{\operatorname{ind}(K, x_i) + \operatorname{ind}(K, x_j)} \Big) = \frac{1}{2} \sum_i (-1)^{\operatorname{ind}(K, x_i)} \sum_{j \neq i} (-1)^{\operatorname{ind}(K, x_j)} \\ &= \frac{1}{2} \sum_i (-1)^{\operatorname{ind}(K, x_i)} \Big(\sum_j (-1)^{\operatorname{ind}(K, x_j)} - (-1)^{\operatorname{ind}(K, x_i)} \Big) \\ &= \frac{1}{2} \big(1 - \operatorname{card}(\{\nabla K = 0\} \cap \{\Delta K < 0\})\big), \end{aligned}$$

a contradiction to our assumptions. Here, the last equality follows from the fact that we are assuming $(*)$ not to hold.

Remark 8.13. The Poincaré–Hopf theorem could have been used in a similar manner for the proof of Theorem C, due to the relation between the index of a critical point of $J_{K,\tau}$ and the local Leray–Schauder degree of the corresponding solution of $(\mathrm{E}_{K,\tau})$.

Chapter 9

Forbidden curvatures

In this chapter we prove non-existence results on $\mathbb{S}^n$ for some proper choices of the prescribed curvature function K. We first introduce the classical examples by Kazdan–Warner, and then present some more recent results from [208]. We then describe some new examples following [166], which show the sharpness of the assumptions in Theorem D.

9.1 First non-existence results

Let us first recall the Kazdan–Warner identity on $(\mathbb{S}^n, g_{\mathbb{S}^n})$, proven in Section A.2 below: if $u > 0$ is a solution to

$$-c_n \Delta_g u + R_g u = K u^{\frac{n+2}{n-2}} \quad \text{on } \mathbb{S}^n \tag{$\mathrm{E_K}$}$$

for some $K\colon \mathbb{S}^n \to \mathbb{R}$, and $y^1, \dots, y^{n+1}$ are Euclidean coordinates of $\mathbb{R}^{n+1} \supseteq \mathbb{S}^n$ restricted to $\mathbb{S}^n$, then

$$\int_{\mathbb{S}^n} \langle \nabla_{g_{\mathbb{S}^n}} K, \nabla_{g_{\mathbb{S}^n}} y^i \rangle u^{\frac{2n}{n-2}} \, dV_{g_{\mathbb{S}^n}} = 0 \quad \text{for every } i = 1, \dots, n+1.$$

Theorem E. *If $K\colon \mathbb{S}^n \to \mathbb{R}$ is positive and monotone with respect to some Euclidean coordinate (for example, if K is affine), then* ($\mathrm{E_K}$) *is not solvable.*

More examples of forbidden functions were given in [36] via the action of conformal vector fields. We present next another result that applies to perturbations of the above monotone functions, proven via blow-up analysis, following an argument from [208] regarding the two-dimensional problem.

Theorem F ([166]). *Suppose that $n = 3$ or 4. Then there exists $K\colon \mathbb{S}^n \to \mathbb{R}$ positive such that $\langle \nabla_{g_{\mathbb{S}^n}} K, \nabla_{g_{\mathbb{S}^n}} y \rangle$ changes sign for every y linear in $\mathbb{R}^{n+1}$ and such that* ($\mathrm{E_K}$) *has no solutions.*

Proof. We argue by contradiction, and take $K = K_j \to \bar{K} > 0$ in $C^2(\mathbb{S}^n)$, for K_j functions with two maxima near the north pole (see Figure 9.1 (i)), $\bar{K}$ a forbidden function (monotone non-decreasing in y^{n+1}), and such that ($\mathrm{E_K}$) with $K = K_j$ is solvable. One can also assume that there exists $\delta > 0$ such that $\Delta K_j \leq -\delta$ in $B_\delta(P_N)$. We also require that the only critical points of $\bar{K}$ are the poles P_N and P_S, and that

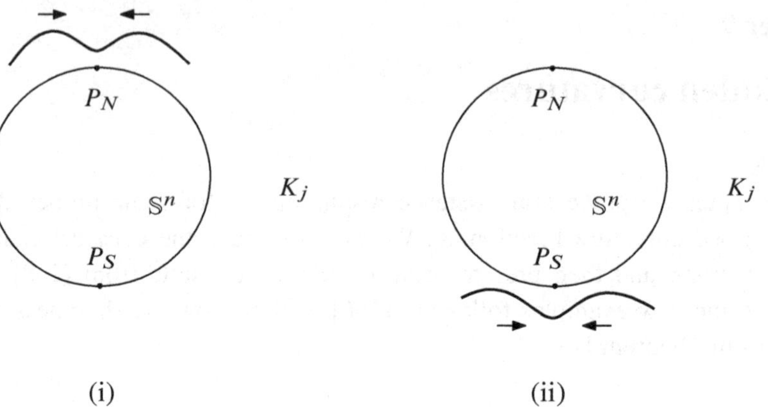

Figure 9.1. Profiles of forbidden curvatures

$\Delta \bar{K}(P_S) > 0$; we know that $\Delta \bar{K}(P_N) < 0$ by the fact that $K_j \to \bar{K}$ in $C^2(\mathbb{S}^n)$. Take for example, in stereographic coordinates from the south pole,

$$K_j(x^1, x^n) = 2 - |x^1|^2 + \tfrac{1}{j}(x^n)^2 - (x^n)^4, \quad \mathbb{R}^{n-1} \times \mathbb{R} \ni (x^1, x^n) \approx (0,0).$$

It is easy to show that such functions are not monotone with respect to any Euclidean coordinate restricted to $\mathbb{S}^n$.

Suppose u_j solves ($\mathrm{E_K}$) with $K = K_j$. Then u_j should blow up as $j \to \infty$. If not, by Ascoli's theorem and elliptic regularity, u_j must sub-converge to a solution of ($\mathrm{E_K}$) with $K = \bar{K}$, which is impossible as $\bar{K}$ is forbidden by Theorem E.

Since $(K_j)_j$ is bounded in $C^2(\mathbb{S}^n)$, we get that $K_j \geq C^{-1}$ for some $C > 0$, and since $n = 3, 4$, $u_j \rightharpoonup 0$ and u_j has only isolated blow-ups (only one if $n = 3$) by Theorem 6.9. If $x_j \to \bar{x}$ is such a blow-up point, we know that $\nabla K_j(x_j) \to 0$. Therefore, $\nabla \bar{K}(\bar{x}) = 0$, and hence either $\bar{x} = P_S$ or $\bar{x} = P_N$. Recall that by Remark 6.11,

$$u_j(x_j)u_j(x) \to \sum_{\ell=1}^{L} \alpha_\ell G_{\bar{x}^\ell}(x),$$

where we sum over blow-up points, and that the Pohozaev identity implies (see (6.12))

$$\Delta \bar{K}(\bar{x}^1)\lambda_j^{-2} + \tau = -\Big(\alpha_1 A(\bar{x}^1) + \sum_{\ell=2}^{L} \alpha_\ell G_{\bar{x}^\ell}(\bar{x}^1)\Big)\lambda_j^{2-n} + \text{l.o.t.}$$

for $j \to +\infty$. As this forbids $\Delta K(\bar{x}) > 0$, it must be $\bar{x} = P_N$, and in particular we only have one blow-up point by the isolated simple property. On the other hand, we are taking $\tau = 0$, so again by Pohozaev's identity we cannot have $\Delta \bar{K}(\bar{x}) < 0$. Therefore, blow-up of the u_j is excluded. Hence, (E_K) with $K = K_j$ has no solution for large j. ■

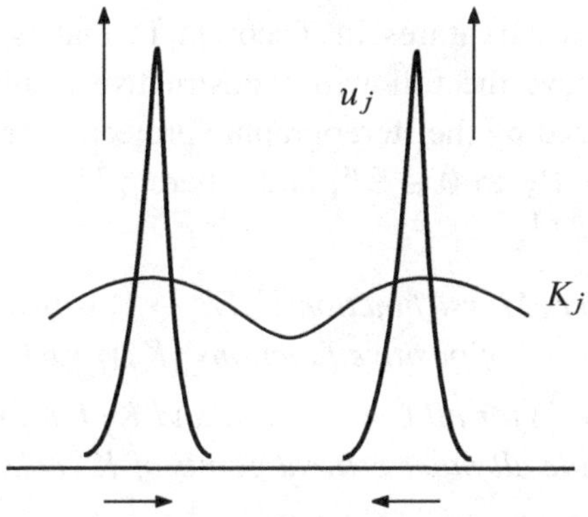

Figure 9.2. Non-isolated blow-ups

Remark 9.1. The above argument does not apply in dimension $n \geq 5$ (see Figure 9.2). In fact, $\bar{K}$ could satisfy the pinching condition (P2) and K_j have two critical points with negative Laplacian. In that case we have existence of solutions to (E_K) with $K = K_j$.

Indeed, blow-ups can be non-isolated in dimension $n \geq 5$. In [148] for the above K_j, solutions with two collapsing bubbles were constructed for $n = 6, 7, 8, 9$ via implicit function arguments; see also [49].

9.2 Forbidden functions with a given Morse structure

We next show more examples and the sharpness of condition (ii) in Theorem D. For a Morse function K on $\mathbb{S}^n$, let

$$\mathcal{M}_j(K) := \#\{x \in \mathbb{S}^n : \nabla K(x) = 0, \ \operatorname{ind}(K, x) = j\},$$

where $\#A$ denotes the cardinality of A.

Theorem G ([166]). *For $n \geq 3$ and any Morse function $\widetilde{K}: \mathbb{S}^n \to \mathbb{R}$ with only one maximum point, there exists $K: \mathbb{S}^n \to \mathbb{R}$ Morse and positive such that*

(i) $\mathcal{M}_\ell(K) = \mathcal{M}_\ell(\widetilde{K})$ *for* $\ell = 0, 1, \ldots, n$;

(ii) *K has only one critical point with negative Laplacian (the maximum);*

(iii) ($\mathrm{E_K}$) *has no solution.*

Moreover, K can be chosen with $\frac{K_{\max}}{K_{\min}}$ arbitrarily close to 1.

We will choose functions $K = K_j$ that are *reversed* (see Figure 9.1 (ii)) with respect to the prescribed curvatures in Theorem F, that is, for $\mathcal{M}_0(\widetilde{K}) = 2$ and $\mathcal{M}_1(\widetilde{K}) = 1$. First we have the following constructive result, where $\{y^1, \ldots, y^n\}$ denote coordinates induced by the stereographic projection from the north pole P_N, mapping the south pole P_S to $0 \in \mathbb{R}^n$, and where y^{n+1} stands for the $(n+1)$th ambient coordinate in $\mathbb{R}^{n+1}$.

Proposition 9.2. *For every Morse function $\widetilde{K}: \mathbb{S}^n \to \mathbb{R}$ with only one local maximum point there exists a sequence of positive functions $(K_j)_j$ such that*

(a) *$\mathcal{M}_\ell(K_j) = \mathcal{M}_\ell(\widetilde{K})$ for all $\ell = 0, \ldots, n$ and K_j has only one local maximum point at P_N, while all other critical points of K_j converge to P_S;*

(b) *there exists a neighborhood U of P_S and $c > 0$ such that $\Delta K_j \geq c$ on U;*

(c) *$K_j \to K_0$ in $C^3(\mathbb{S}^n)$, where K_0 is a positive monotone non-decreasing function in y^{n+1}, affine and non-constant in y^{n+1} outside a small neighborhood of P_S.*

Proof. We proceed with seven technical steps.

Step 1. For $\delta_0 > 0$, $\varepsilon_0 > 0$ small, take a function $\mathcal{K}$ such that

$$\begin{cases} \mathcal{K} = \dfrac{\varepsilon_0}{8n^4}(y^n)^2 & \text{for } y^{n+1} \leq -1 + \delta_0, \\ \mathcal{K} = \varepsilon_0(1 + y^{n+1}) & \text{for } y^{n+1} \geq -1 + 2\delta_0, \\ \langle \nabla \mathcal{K}, \nabla y^{n+1} \rangle \geq 0 & \text{on } \mathbb{S}^n \setminus \{P_N, P_S\}. \end{cases}$$

We can also suppose that

$$\{\nabla \mathcal{K} = 0\} \cap \{y^{n+1} \leq -1 + 2\delta_0\} \subseteq \{y^{n+1} \leq -1 + \delta_0\}.$$

The function $\mathcal{K}$ can be chosen so that its Laplacian with respect to the y-coordinates is bounded away from zero in $\{y^{n+1} \geq -1 + 2\delta_0\}$. If $\varphi_{\pi_{P_N}}$ is the conformal factor of π_{P_N}, i.e. $g_{\mathbb{S}^n} = \varphi_{\pi_{P_N}} dy^2$, then

$$\Delta_{g_{\mathbb{S}^n}} \mathcal{K} = \varphi_{\pi_{P_N}}^{-1} \Delta_{g_{\mathbb{R}^n}} \mathcal{K} + O(|\nabla \varphi_{\pi_{P_N}}| |\nabla \mathcal{K}|).$$

Therefore, $\mathcal{K}$ satisfies $\Delta_{g_{\mathbb{S}^n}} \mathcal{K} \geq c > 0$ on $U := \{y^{n+1} < -1 + 2\delta_0\}$.

Step 2. We consider next a Morse function $\widetilde{K}$ with given numbers of critical points of fixed indices and only one local maximum, which we can suppose to coincide with P_N. We compose $\widetilde{K}$ with a Möbius map Φ preserving P_N, so that all other critical points $\{p_1, \dots, p_l\}$ of $\widetilde{K} \circ \Phi$ lie in the set $\{y^{n+1} \le -1 + \frac{1}{4}\delta_0\}$, where δ_0 is as in the previous step. Composing with Φ does not affect either the non-degeneracy or the index of critical points of the function $\widetilde{K}$.

Step 3. For δ_0 small, the coordinates of the points p_i, still denoted by p_i, are of the form

$$p_i = (p_i', p_i^n) \quad \text{with } p_i' \in \mathbb{R}^{n-1}, p_i^n \in \mathbb{R}, \text{and } (p_i', p_i^n) \in B_{\delta_0^{1/4}}(0) \subseteq \mathbb{R}^n.$$

Up to some rotation around $0 \in \mathbb{R}^n$, we may assume that $p_i' \neq p_l' \in \mathbb{R}^{n-1}$ for $i \neq l$.

Step 4. Since $\widetilde{K} \circ \Phi$ is Morse, there exist $R_i \in \mathrm{SO}(n)$ and a diagonal non-singular matrix A_i such that, near p_i,

$$(\widetilde{K} \circ \Phi)(y) = \langle R_i(y - p_i), A_i R_i(y - p_i)\rangle + O(|y - p_i|^3).$$

Without affecting the Morse structure of $\widetilde{K}$, we can modify it so that for some $\delta_1 \ll \delta_0$ one has precisely

$$(\widetilde{K} \circ \Phi)(y) = \langle R_i[y - p_i], A_i R_i[y - p_i]\rangle \quad \text{for } |y - p_i| \le \delta_1.$$

Since no p_i is a local maximum, we can also assume that the last diagonal entry of A_i is positive.

Step 5. We next consider a smooth curve $\gamma_i \colon [0,1] \to \mathrm{SO}(n)$ such that

$$\gamma_i(0) = \mathrm{Id}_n \quad \text{and} \quad \gamma_i(1) = R_i,$$

and then introduce the new function

$$\Theta_i(y) := \langle \gamma_i(f(|y - p_i|^2))[y - p_i] A_i \gamma_i(f(|y - p_i|^2))[y - p_i]\rangle, \quad |y - p_i| \le \delta_1,$$

where f is 0 in a neighborhood of 0 and coinciding with 1 in a neighborhood of δ_1^2. We claim that p_i is the only critical point of this function in $B_{\delta_1}(p_i)$. Indeed, consider a curve in $\mathbb{R}^n$ of the type

$$Y_t := p_i + t\big(\gamma_i(f(t^2))\big)^{-1} Y \quad \text{with } Y \in \mathbb{R}^n, |Y| = 1, \text{and for } t \in [0, \delta_1].$$

Then clearly $\Theta_i(Y_t) = t^2 \langle Y, AY\rangle$, so whenever $\langle Y, AY\rangle \neq 0$ the gradient of Θ_i is non-zero for $t \neq 0$. On the other hand, if $\langle Y, AY\rangle = 0$, one can always consider a trajectory Y_s in the unit sphere such that

$$\frac{d}{ds}\Big|_{s=0} \langle Y_s, AY_s\rangle \neq 0.$$

If Y_t is as in the previous formula, consider the curve $Y_t(s)$ replacing Y with $Y(s)$. Then its s-derivative is a non-critical direction for Θ_i. In this way we have proved that

$$\exists 0 < \delta_2 \ll \delta_1 \colon \Theta_i(y) = \langle y - p_i, A_i[y - p_i]\rangle \quad \text{for } |y - p_i| \le \delta_2,$$

with A_i diagonal and $(A_i)_{nn} > 0$. Replacing $\widetilde{K} \circ \Phi$ with Θ_i near each p_i, no further critical point is created and the Morse structure is preserved.

Step 6. Recall that we performed a rotation of coordinates so that the first $n - 1$ components of the points p_i, i.e. $p_1', \dots, p_l' \in \mathbb{R}^{n-1}$ are all distinct. Then there exists a constant $0 < \delta_3 \ll \delta_2$ such that $|p_i' - p_j'| \ge 4\delta_3$ if $i \ne j$. Choose next a cutoff function $\mathcal{G}$ so that

$$\begin{cases} \mathcal{G} = p_i^n & \text{on } B_{\delta_3}(p_i), \\ \mathcal{G} = 0 & \text{on } \mathbb{R}^{n-1} \setminus \bigcup_{i=1}^{l} B_{2\delta_3}(p_i). \end{cases}$$

Denote by Θ the function obtained on replacing $\widetilde{K} \circ \Phi$ by the above Θ_i near each p_i, and define

$$\widetilde{\Theta}(y', y^n) = \Theta(y', y^n + \mathcal{G}(y')).$$

Then the only critical points of $\widetilde{\Theta}$ are $(p_1', 0), \dots, (p_l', 0)$. Indeed, these are critical points by construction, and

$$\begin{cases} \nabla_{y'}\widetilde{\Theta}(y', y^n) = \nabla_{y'}\Theta(y', y^n + \mathcal{G}(y')) \\ \qquad\qquad - \partial_{y^n}\Theta(y', y^n + \mathcal{G}(y'))\nabla_{y'}\mathcal{G}(y'), \\ \partial_{y^n}\widetilde{\Theta}(y', y^n) = \partial_{y^n}\Theta(y', y^n + \mathcal{G}(y')). \end{cases}$$

This implies that $\nabla\widetilde{\Theta}(y', y^n) = 0$ if and only if $\nabla\Theta(y', y^n + \mathcal{G}(y')) = 0$, namely the desired conclusion.

Final step. Let $\widehat{K}$ be the function obtained from $\widetilde{K}$ following the previous steps and consider a sequence of Möbius maps Φ_j fixing P_N and sending every other point to P_S as $j \to \infty$. Starting from a Morse function $\widetilde{K}$ as in the statement, apply Step 3 to Step 6. For ε_0 small fixed and $\varepsilon_j \searrow 0$ consider a function K_j of the form ($\widehat{K} = \widehat{K}_j$)

$$K_j = 1 + \varepsilon_0 \mathcal{K} + \varepsilon_j \widehat{K}.$$

Using the fact that $\mathcal{K} \equiv 0$ for $y^n = 0$ and $|y|$ small, one can check that every critical point of K_j is either P_N, the global maximum, or converges to P_S with $\mathcal{M}_j(K_j) = \mathcal{M}_j(\widetilde{K})$ for all j. If we let $\varepsilon_j \searrow 0$ sufficiently fast, then K_j satisfies the desired properties choosing $K_0 = 1 + \varepsilon_0 \mathcal{K}$. ■

9.3 Blow-up analysis via study of singular solutions

Here we prove Theorem G arguing by contradiction, assuming that (E_K) for $K = K_j$ is solvable for all j, with K_j converging to $\bar{K}$. We will use blow-up analysis, but differently from Chapter 6, to derive a final contradiction. We begin with the following lemma, whose proof is somehow related to the argument in Remark 6.11 (ii).

Lemma 9.3 ([62]). *Let u_j be positive solutions of* (E_K) *with $K = K_j$. Then the blow-up points of u_j can only be where $\nabla \bar{K} = 0$, namely at P_N or P_S.*

We can next exclude the south pole as a blow-up point.

Lemma 9.4. *Let u_j be solutions of* (E_K) *with $K = K_j$. Then P_S cannot be a blow-up point for $(u_j)_j$.*

Proof. We only give here a sketch of the proof, referring to [166] for details, which consists in showing first that blow-ups are isolated simple, along the lines of the proof of Theorem 6.9. For $n = 3, 4$ this result can be directly applied. We therefore assume from now on that $n \geq 5$: the proof works in all dimensions, due to the fact that near P_S one has $\Delta K_j \geq \delta > 0$, which favors deconcentration of bubbles, and so does the presence of other nearby bubbles. If $u_j \approx \varphi_{p_j,\lambda_j}$, with $p_j \to P_S$ and $\lambda_j \to \infty$, then we can use Pohozaev's identity (6.12) to find that, up to positive coefficients in each term (and for $r > 0$ small),

$$\frac{\Delta K_j(p_j)}{\lambda_j^2} = \int_{\partial B_r} \mathcal{B}\, d\sigma = O(u_j(p_j)^{-2}) = O(\lambda_j^{-(n-2)}).$$

Since $\Delta K_j(p_j) \geq \delta > 0$ and $n \geq 5$, we get a contradiction. ∎

Lemma 9.5. *If u_j is as above, then $u_j \to 0$ in $C^2_{\mathrm{loc}}(\mathbb{S}^n \setminus \{P_N\})$.*

Proof. Since $(u_j)_j$ is uniformly bounded in compact sets of $\mathbb{S}^n \setminus \{P_N\}$ and $K_j \to \bar{K}$ in $C^3(\mathbb{S}^n)$, we must have $u_j \to \bar{u} \geq 0$ in $C^2_{\mathrm{loc}}(\mathbb{S}^n \setminus \{P_N\})$. Moreover, $\bar{u}$ would solve (E_K) with $K = \bar{K}$ in $\mathbb{S}^n \setminus \{P_N\}$. We now consider the following two cases:

Case 1 ($\bar{u}$ has a removable singularity at P_N). This is not possible as $\bar{u} > 0$ would solve (E_K) on $\mathbb{S}^n$ with $K = \bar{K}$, a forbidden curvature.

Case 2 ($\bar{u}$ has a non-removable singularity at P_N). In this case, by a result in [219], $\bar{u}$ is *asymptotically radial* near P_N. Moreover, after rescaling, the profile of $\bar{u}$ near P_N must be a (possibly singular) solution of the Yamabe equation in $\mathbb{R}^n \setminus \{0\}$. Such solutions were classified by the following theorem.

Theorem 9.6 ([47]). *Suppose $u_0 > 0$ solves*

$$-\Delta u_0 = u_0^{\frac{n+2}{n-2}} \quad \text{on } \mathbb{R}^n \setminus \{0\}.$$

Then either u_0 has a removable singularity (and therefore it is as in Theorem 5.1*), or u_0 is radial and of the form*

$$u_0(x) = |x|^{\frac{2-n}{2}} v(\log |x|),$$

with v positive and periodic in its argument (see Figure 9.3*).*

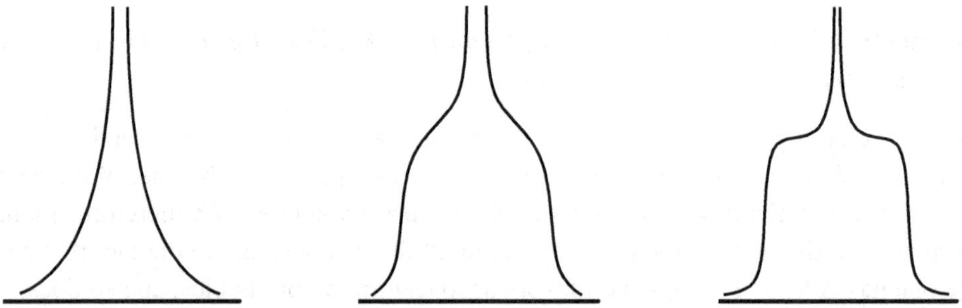

Figure 9.3. Fowler's singular solutions

Such solutions, called *Fowler's solutions*, are counterparts of Delaunay's axially symmetric constant mean curvature (CMC) surfaces, and can be classified using ODE analysis.

We can now use the Kazdan–Warner identity (similar to Pohozaev's) for $\bar{u}$ on $\mathbb{S}^n \setminus B_r(P_N)$ to derive a contradiction. As for the proof of Proposition A.4 below, one finds the identity

$$\int_{\mathbb{S}^n \setminus B_r(P_N)} \langle \nabla_{g_{\mathbb{S}^n}} \bar{K}, \nabla_{\mathbb{S}^n} y^{n+1} \rangle \bar{u}^{2^*} \, dV_{g_{\mathbb{S}^n}} + \int_{\partial B_r(P_N)} \mathcal{B}(\bar{u}) \, d\sigma = 0,$$

where in the boundary term we used the conformal change induced by stereographic coordinates, while keeping the same notation for $\bar{u}$.

However, given the above asymptotic radiality of $\bar{u}$ and the classification from Theorem 9.6, for $r > 0$ small it can be proved that

$$\int_{\partial B_r(P_N)} \mathcal{B}(\bar{u}) \, d\sigma = \int_{\partial B_r(P_N)} \mathcal{B}(u_0) \, d\sigma + o_r(1) > 0,$$

where the latter inequality follows from the explicit form of the radial profile v, studied in [108]. This excludes the second case as well.

Case 3 ($\bar{u} \equiv 0$). This is the only possible case left, which concludes the proof. ■

Next, even the latter case in the former proof will be excluded, with a further blow-up analysis and new arguments.

Lemma 9.7. *There exists a blow-down v_j of u_j near P_N,*

$$v_j(x) := \mu_j^{\frac{n-2}{2}} u_j(\mu_j x), \quad \mu_j \to 0,$$

which has a non-zero limit in $C^2_{\mathrm{loc}}(\mathbb{R}^n \setminus \{0\})$.

Proof. Choose stereographic coordinates x near P_N using π_S, so $x = 0$ represents the point P_N. We blow up $g_{\mathbb{S}^n}$ conformally near the north pole P_N in order to obtain a metric

$$\tilde{g} = \tilde{u}^{\frac{4}{n-2}} g_{\mathbb{S}^n}, \quad \tilde{u} \approx |x|^{\frac{2-n}{2}} \text{ near } x = 0,$$

in the above coordinates, and with a cylindrical end of bounded geometry, see Figure 9.4.

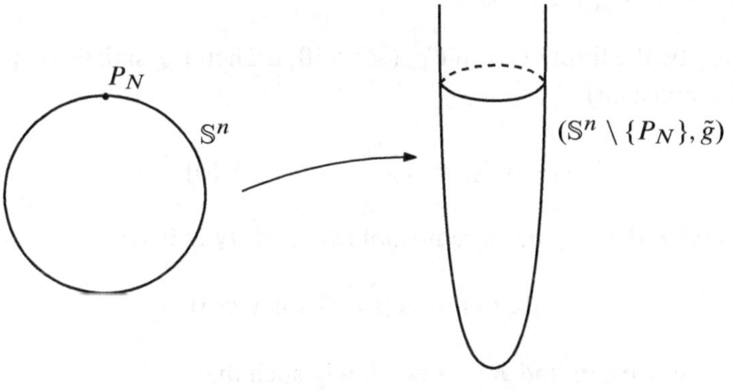

Figure 9.4. Cylindrical conformal blow-up

The function $\tilde{u}_j := \tilde{u}^{-1} u_j$ is the conformal representative of u_j in this new metric. By a result in [63], there exists $C > 0$ such that

$$u_j(x) \le C|x|^{\frac{2-n}{2}},$$

and consequently, $(\tilde{u}_j)_j$ is uniformly bounded. Notice that this yields $v_j \le C|x|^{\frac{2-n}{2}}$ for all choices of small numbers $\mu_j > 0$.

We now assume by contradiction that

$$v_j \to 0 \text{ in } C^2_{\mathrm{loc}}(\mathbb{R}^n \setminus \{0\}) \text{ for every choice of } \mu_j \to 0.$$

Then $\tilde{u}_j \to 0$ locally uniformly in $\mathbb{S}^n \setminus \{P_N\}$. Moreover, as $\tilde{u}_j$ satisfies

$$L_{\tilde{g}} \tilde{u}_j = K_j \tilde{u}_j^{\frac{n+2}{n-2}} \quad \text{on } (\mathbb{S}^n \setminus \{0\}, \tilde{g}),$$

by standard elliptic estimates one finds that

$$\|\tilde{u}_j\|_{L^\infty(B_1(x))} \le C\|\tilde{u}_j\|_{L^\infty(B_1(x))}^{\frac{n+2}{n-2}},$$

for x in the cylindrical end of $(\mathbb{S}^n \setminus \{P_N\}, \tilde{g})$, where the metric ball $B_1(x)$ is with respect to the metric $\tilde{g}$. Since $\|\tilde{u}_j\|_{L^\infty(B_1(x))} \to 0$, we must have $\|\tilde{u}_j\|_{L^\infty(B_1(x))} = 0$ for j large, a contradiction to the positivity of u_j. ■

We next perform a blow-down analysis at a *slowest possible rate* as follows. Due to Lemma 9.7, we can choose $\bar{\mu}_j \to 0$ with the following properties:

(i) $\bar{v}_j(x) := \bar{\mu}_j^{\frac{n-2}{2}} u_j(\bar{\mu}_j x)$ converges in $C^2_{\mathrm{loc}}(\mathbb{R}^n \setminus \{0\})$ to a non-zero limit.

(ii) If $\frac{\hat{\mu}_j}{\bar{\mu}_j} \to 0$, then $\hat{v}_j(x) := \hat{\mu}_j^{\frac{n-2}{2}} u_j(\hat{\mu}_j x) \to 0$ in $C^2_{\mathrm{loc}}(\mathbb{R}^n \setminus \{0\})$.

Lemma 9.8. *Let $\bar{\mu}_j$ and $\bar{v}_j$ be as above. Then, up to a subsequence, $\bar{v}_j$ converges to a regular bubble in $C^2_{\mathrm{loc}}(\mathbb{R}^n \setminus \{0\})$.*

Proof. Let $\bar{v}_\infty$ be the limit of $\bar{v}_j$ in $C^2_{\mathrm{loc}}(\mathbb{R}^n \setminus \{0\})$. Then $\bar{v}_\infty$ satisfies (up to a positive multiplicative constant)

$$-\Delta \bar{v}_\infty = \bar{v}_\infty^{\frac{n+2}{n-2}} \quad \text{on } \mathbb{R}^n \setminus \{0\}.$$

By Theorem 9.6 either $\bar{v}_\infty$ has a removable singularity at 0, or

$$\bar{v}_\infty(x) \approx |x|^{\frac{2-n}{2}} \quad \text{for } x \simeq 0.$$

In the latter case, we can find $R_j \to \infty$ slowly such that

$$\frac{1}{C}|x|^{\frac{2-n}{2}} \le v_j(x) \le C|x|^{\frac{2-n}{2}} \quad \text{for } 1 \le |x| \le R_j^2.$$

Then, with

$$\hat{v}_j(x) := \hat{\mu}_j^{\frac{n-2}{2}} u_j(\hat{\mu}_j x) = R_j^{\frac{n-2}{2}} \bar{v}_j(R_j x), \quad \hat{\mu}_j := R_j \bar{\mu}_j,$$

we see that $\hat{v}_j$ converges to a non-zero limit in $C^2_{\mathrm{loc}}(\mathbb{R}^n \setminus \{0\})$, which violates condition (ii) above on $\bar{\mu}_j$. ■

Consider next a Kelvin inversion around a sphere of radius $\check{\mu}_j \to 0$ with $\frac{\check{\mu}_j}{\bar{\mu}_j} \to 0$. In the coordinates given by the stereographic projection π_{P_S}, this corresponds to the map $x \mapsto \frac{\check{\mu}_j^2 x}{|x|^2}$. Letting

$$\check{u}_j(x) := \frac{\check{\mu}_j^{n-2}}{|x|^{n-2}} u_j\Big(\frac{\check{\mu}_j^2 x}{|x|^2}\Big), \quad \check{K}_j(x) := K_j\Big(\frac{\check{\mu}_j^2 x}{|x|^2}\Big),$$

we obtain

$$-c_n \Delta \check{u}_j = \check{K}_j \check{u}_j^{\frac{n+2}{n-2}}.$$

The function $\check{K}_j$ is highly oscillating near $x = 0$, and it has the expansion

$$\check{K}_j(x) = \bar{K}(0) - c_0(1 + o_j(1))\frac{\check{\mu}_j^4}{|x|^2} + O(\check{\mu}_j^6 |x|^{-3}) \quad \text{for } |x| \geq \frac{\check{\mu}_j^2}{\delta},$$

for some $c_0 > 0$. Let U_0 be a radial solution of $-\Delta U_0 = U_0^{\frac{n+2}{n-2}}$ in $\mathbb{R}^n$, and set

$$U_{\lambda,a}(x) := \lambda^{\frac{n-2}{2}} U_0(\lambda(x-a)), \quad a \in \mathbb{R}^n,\ \lambda > 0.$$

By Lemma 9.8, u_j is close in H^1 to a multiple (depending on $\bar{K}(0)$) of U_{λ_j,a_j} with $\lambda_j \approx \bar{\mu}_j^{-1}$ on a proper annulus centered at $x = 0$. Since, as remarked before,

$$u_j(x) \leq C|x|^{\frac{2-n}{2}},$$

it follows that $\lambda_j|a_j| \leq C$. If $U_{\check{\lambda}_j,\check{a}}$ is the corresponding Kelvin inversion of U_{λ_j,a_j}, then

$$\check{a}_j = \lambda_j^2 \check{\mu}_j^2 \frac{a_j}{1 + \lambda_j^2|a_j|^2}, \quad \check{\lambda}_j = \frac{1 + \lambda_j^2|a_j|^2}{\lambda_j \check{\mu}_j^2},$$

which gives $\check{\lambda}_j|\check{a}_j| = \lambda_j|a_j| \leq C$. Hence, there exists $x_j \to 0$ such that

$$\check{u}_j(x_j) \approx \left(\frac{\check{\mu}_j^2}{\bar{\mu}_j}\right)^{\frac{2-n}{2}} \to \infty,$$

and $\check{u}_j$ develops a bubble at a scale $\frac{\check{\mu}_j^2}{\bar{\mu}_j} \to 0$. Notice that $\check{u}_j(x) \leq C|x|^{\frac{2-n}{2}}$, and hence $x = 0$ is the only blow-up point for $(\check{u}_j)$. Moreover, $\max \check{u}_j \approx (\frac{\check{\mu}_j^2}{\bar{\mu}_j})^{(2-n)/2}$. Due to the regular bubble profile of $(u_j)_j$ as in Lemma 9.8, one gets that the function

$$s \mapsto \bar{\bar{w}}_j(s) = s^{\frac{n-2}{2}} \frac{1}{|\partial B_s|} \int_{\partial B_s(x_j)} \check{u}_j \, d\sigma,$$

has a unique critical point for s of order $\frac{\check{\mu}_j^2}{\bar{\mu}_j}$. If there is another critical point at some $\check{r}_j \to 0$, then it must be at $\check{r}_j \gg \frac{\check{\mu}_j^2}{\bar{\mu}_j}$. Therefore, we can choose $\check{\mu}_j$ so that $\bar{\bar{w}}_j$ has a unique critical point in $[\frac{\check{\mu}_j^2}{\bar{\mu}_j}, 1]$. For such a choice of $\check{\mu}_j$ one has for some $0 < r \leq 1$ (following the proof of Proposition 6.7),

$$\begin{aligned} \check{u}_j(x) &\leq C\check{u}_j(x_j)^{-1}|x - x_j|^{2-n} && \text{on } B_r, \\ \check{u}_j(x_j)\check{u}_j(x) &\to a|x|^{2-n} + b(x) && \text{in } C^2_{\text{loc}}(B_r \setminus \{0\}), \end{aligned}$$

for some $a > 0$ and a harmonic function b.

Finally we prove our non-existence result, which follows from the next proposition.

Proposition 9.9. *Problem* ($\mathrm{E_K}$) *with* $K = K_j$ *has no positive solution for* j *large.*

Proof. We assume by contradiction that ($\mathrm{E_K}$) with $K = K_j$ has a positive solution for every j. For $n = 3, 4$ this goes as for the proof of Theorem F, so we assume that $n \geq 5$. Then, with the same notation as before, using that

$$\check{u}_j(x_j)\check{u}_j(x) \to a|x|^{2-n} + b(x) \quad \text{in } C^2_{\text{loc}}(B_r \setminus \{0\}),$$

and

$$\check{u}_j \leq C\Big(\frac{\check{\mu}_j^2}{\bar{\mu}_j}\Big)^{\frac{2-n}{2}}, \quad |x| \leq \frac{\check{\mu}_j^2}{8},$$

one finds from $n \geq 5$ that

$$\begin{aligned}
\int_{\partial B_r} \check{K}_j u_j^{2^*}\, d\sigma &= o(\max \check{u}_j)^{-2} = o\Big(\frac{\check{\mu}_j^2}{\bar{\mu}_j}\Big)^2, \\
\int_{\partial B_r} \mathcal{B}\, d\sigma &= O(\max \check{u}_j)^{-2} = o\Big(\frac{\check{\mu}_j^2}{\bar{\mu}_j}\Big)^2.
\end{aligned}$$

Moreover, as $\check{u}_j$ develops a bubble at scale $\frac{\check{\mu}_j^2}{\bar{\mu}_j}$, we have

$$\int_{B_{\check{\mu}_j^2/8}} \langle x, \nabla \check{K}_j\rangle \check{u}_j^{2^*}\, dx = O(\bar{\mu}_j^n), \quad \int_{B_r \setminus B_{\check{\mu}_j^2/8}} \langle x, \nabla \check{K}_j\rangle \check{u}_j^{2^*}\, dx \geq c\bar{\mu}_j^2,$$

for some $c > 0$. This leads to a contradiction to the Pohozaev identity; see (6.12) and Corollary A.2. ∎

Chapter 10

Related directions and perspectives

In this chapter we collect some results and perspectives on some topics related to those presented above, although not central to the scope of these notes. We cannot enter here into much detail, but for each topic we will give a list of references which, although certainly incomplete, we hope could still be useful to the interested reader.

10.1 Blow-up analysis in general dimensions

On a closed manifold (M^n, g) of positive Yamabe class ($\lambda_1(L_g) > 0$), we consider the sub-critical approximation

$$L_g u = -c_n \Delta_g u + R_g u = K u^{\frac{n+2}{n-2}-\tau} \quad \text{for } \tau > 0 \text{ small}, \qquad (\mathrm{E}_{K,\tau})$$

of the prescribed curvature equation

$$-c_n \Delta_g u + R_g u = K u^{\frac{n+2}{n-2}}. \qquad (\mathrm{E}_K)$$

Recall that, by the result in Chapter 6, for $n = 3, 4$, blow-ups of solutions to $(\mathrm{E}_{K,\tau})$ as $\tau \searrow 0$ are isolated simple, even for K varying with τ (with uniform positive upper and lower bounds, as well as with uniform regularity estimates). On $(\mathbb{S}^n, g_{\mathbb{S}^n})$ and K Morse (fixed), we used the following theorem in Chapter 9.

Theorem 10.1 ([63]). *Let $(M^n, g) = (\mathbb{S}^n, g_{\mathbb{S}^n})$. Suppose that K is Morse, and that $\bar{x} \in \mathbb{S}^n$ satisfies $\nabla K(\bar{x}) = 0$, $\Delta K(\bar{x}) < 0$. Suppose $\tau_i \searrow 0$, and that $(u_i)_i$ solve $(\mathrm{E}_{K,\tau})$ with $\tau = \tau_i$ and blow up at $\bar{x}$. Then there exists $C, r > 0$ such that*

$$u_i(x) \le C \operatorname{dist}(x, \bar{x})^{\frac{2-n}{2}}$$

for $\operatorname{dist}(x, \bar{x}) \le r$. In particular, the blow-up is isolated.

Still in [63], the authors obtained more specialized results for the case $\tau = 0$, i.e. for the original problem (E_K), as described in the next theorem.

Theorem 10.2 (Chen–Lin [63]). *Let $(M^n, g) = (\mathbb{S}^n, g_{\mathbb{S}^n})$, and suppose that K is positive, Morse, and it satisfies* (ND). *Then*

(i) *for $n = 5$, solutions of (E_K) are uniformly bounded;*

(ii) *for $n = 6$, in the case of blow-up of a sequence of solutions $(u_i)_i$ of (E_K), one has that $\|u_i\|_{H^1(\mathbb{S}^n)} \le C$ and the weak limit of $(u_i)_i$ is non-zero;*

(iii) *for $n \geq 7$, in the case of blow-up of a sequence of solutions $(u_i)_i$ to* (E_K), *one has $\|u_i\|_{H^1(\mathbb{S}^n)} \to +\infty$, and indeed also*

$$\int_{\mathbb{S}^n} u_i^{2^*} \, dV_{g_{\mathbb{S}^n}} \xrightarrow{i \to \infty} +\infty.$$

It would be interesting to derive more information of the above type for solutions u_τ of the sub-critical equation ($\mathrm{E}_{K,\tau}$), as well as for $\tau \searrow 0$.

Moreover, while the latter result provides *necessary* conditions on blowing-up solutions, to our knowledge there is at the moment no constructive example of such behavior. Attempting the use of the abstract approach from Chapter 7, one might need to glue together an increasing number of bubbles as τ tends to zero; see e.g. [64, 179] for related results in this spirit, although under different assumptions.

As pointed out in [63], there are relations between sequences of solutions of (E_K) with diverging energy and singular solutions of the Yamabe equation, which were discussed in Chapter 9. See also [146] for a refined analysis of solutions of the Yamabe equation on $\mathbb{S}^n$ with finitely many singularities.

10.2 Morse theory and Morse homology

As we saw in Chapter 8, Morse theory can be useful to study problem (E_K): here we discuss some related approaches. It would be desirable if any of these allowed the pinching conditions in Theorem D to be relaxed or removed.

Morse theory relates the structure of critical points of a given function to the homology of the manifold where it is defined, via the *handle-body* decomposition obtained through the Morse lemma. We recall next the notion of *Morse homology*; see e.g. [203] for details. It is a homological theory of *dynamical type* obtained from *Morse–Smale functions*. The latter are Morse functions such that for each couple of critical points x, y with indices i and $i-1$ respectively, one has that $W_u(x) \cap W_s(y)$ is either empty or a smooth curve. Here, $W_u(x)$ and $W_s(y)$ denote the *unstable manifold* of x and the *stable manifold* of y for the steepest descent flow. As for the standard Morse condition, this is a generic assumption guaranteed by transversality properties.

Such *connecting orbits* allow one to define a boundary operator ∂ from chains of critical points of index i to those of index $i-1$, and such that $\partial^2 = 0$. This induces a homology theory which in many cases is equivalent to the (standard) singular one. However, one can get a more precise description from the global structure of connecting orbits and the fact that generic energy sub-levels deform by retraction to a union of those; see Figure 10.1. Some papers exploiting these ideas for problem (E_K) are

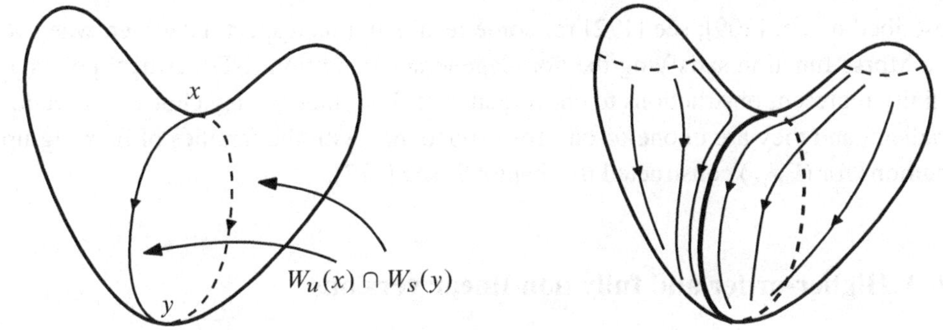

Figure 10.1. Connecting orbits and deformation of sub-levels

[17, 18, 21], under other assumptions that involve some pinching requirement too; see also [174].

There is another dynamical approach to problem (E_K), via the theory of *critical points at infinity*; see [20]. This consists in dealing directly with the functional J_K in (4.3), without introducing a sub-critical approximation, and by analyzing the effect on the topology of sub-levels induced by non-compact flow lines resulting from a steepest descent flow. Such trajectories are expected in general to exist, by Proposition 4.3; see Figure 10.2.

This perspective allows one to understand the geometry of the functional J_K by means of a *Morse lemma at infinity*, according to which, naively, in the handle-body decomposition of sub-levels, non-compact flow lines replace the usual stable manifolds of critical points. This was the strategy employed in [24] to prove Theorem C, and allowed the same authors to solve the Yamabe equation under Dirichlet boundary data in domains with non-trivial topology; see [23], as well as [22] for the geometric case.

Non-compact flow lines are in general delicate objects that might depend on the specific choice of descent flows that are employed, typically *pseudo-gradient flows* as

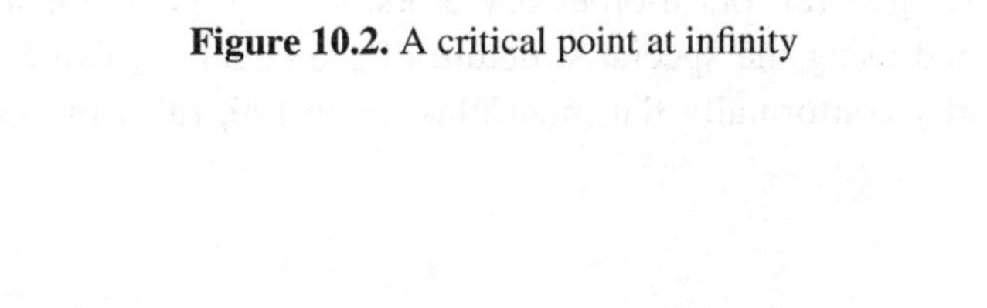

Figure 10.2. A critical point at infinity

described in e.g. [209]; see [172] for some results in this respect. However, when K is a Morse function satisfying the non-degeneracy condition (ND), critical points at infinity represent obstructions to convergence of flow lines for any choice of pseudo-gradient, and they are in one-to-one correspondence with the families of blowing-up solutions for $(\mathrm{E}_{K,\tau})$ constructed in Chapter 7; see [173].

10.3 Higher-order and fully non-linear versions

As discussed in Section 2.2, the Laplace operator in two dimensions and the conformal Laplacian in higher dimensions enjoy natural covariance properties. For $n = 2$, if $\tilde{g} = e^{2w} g$, then $\Delta_{\tilde{g}} = e^{-2w}\Delta_g$; for $n \geq 3$, if $\tilde{g} = u^{\frac{4}{n-2}} g$, then $L_{\tilde{g}}\varphi = u^{-\frac{n+2}{n-2}} L_g(u\varphi)$. In dimension $n \geq 4$, there exists a scalar quantity Q_g and fourth-order operator P_g, known as the *Q-curvature* and the *Paneitz operator* respectively (see [38, 185]), for which analogous features hold.

More precisely, on a Riemannian manifold (M^n, g), for proper dimensional constants $A_n, \dots, E_n$ they are defined as

$$\begin{aligned} Q_g &= A_n \Delta_g R_g + B_n R_g^2 + C_n |\operatorname{Ric}_g|^2, \\ P_g\varphi &= \Delta_g^2 \varphi + \operatorname{div}_g(D_n R_g \nabla\varphi + E_n \operatorname{Ric}_g[\nabla\varphi, \cdot]) \\ &\quad + \frac{n-4}{2} Q_g \varphi, \quad \varphi \in C^\infty(M). \end{aligned}$$

If $n = 4$ and $\tilde{g} = e^{2w} g$, then

$$P_{\tilde{g}} = e^{-4w} P_g \quad \text{and} \quad P_g w + 2Q_g = 2Q_{\tilde{g}} e^{4w},$$

while if $n \geq 5$ and $\tilde{g} = u^{\frac{4}{n-4}}$, then

$$P_{\tilde{g}}\varphi = u^{-\frac{n+4}{n-4}} P_g(u\varphi) \quad \text{and} \quad P_g u = Q_{\tilde{g}} u^{\frac{n+4}{n-4}}.$$

These objects play important roles in the study of 4-manifolds [56, 121], spectral theory [41], and the holography principle in Ads/CFT duality [103, 115].

The problem of prescribing the Q-curvature conformally as a given function has also been studied by several authors; see e.g. [26, 90, 92, 93, 104, 167].

The case of constant Q-curvature has been dealt with in four dimensions in [61, 91, 150], even though some cases still remain open. In higher dimensions, as one is interested in positive solutions, there are also issues with the maximum principle since this might fail for general fourth-order operators; see e.g. [117]. In some situations this can be gained using the special structure of the equation, like for example for Einstein or locally conformally flat manifolds, as in [90, 187]. In positive Yamabe class and, roughly, for metrics of non-negative Q-curvature, a maximum principle for P_g has been obtained by iterating the second-order one; see [122, 126]. In all other cases though, there is at the moment no general progress. We also mention that there exist conformally covariant operators of arbitrary even order (see [114]), which might lead to various types of blow-ups [134], and more recently some fractional versions have also been considered, e.g. in [55, 137, 138, 180, 216].

The above-mentioned paper [41] extended a well-known formula by Polyakov on *functional determinants* from two to four dimensions. These quantities are constructed out of spectra of conformally covariant operators (the Laplacian, in two dimensions) and their extremals within a given conformal class provide a natural choice of representative: for example, in two dimensions extremals correspond to metrics with constant Gaussian curvature; see [182, 183]. In [41], explicit formulas expressing the change of determinants under conformal deformations of the metric were derived, depending on the specific operator through a triple of coefficients. The geometric structure of these functionals is very sensitive to the latter triple, and it leads to quite involved Euler–Lagrange equations with differential terms of different orders and possibly of degenerate type. Some existence (respectively, uniqueness) results, still in limited cases, can be found in [61, 101, 119] (respectively, in [39, 120]). The blow-up analysis for such equations is at rather preliminary stages, and a better understanding of it would likely lead to new existence results. To describe a simple situation, we mention the variation of the functional determinant for the Paneitz operator on the flat four-dimensional torus $(\mathbb{T}^4, dx^2)$, which is given by the formula

$$I(u) := \int_{\mathbb{T}^4} \big[9(\Delta u)^2 + 32\Delta u|\nabla u|^2 + 16|\nabla u|^4\big]\, dx.$$

This functional has a mountain-pass structure, but it lacks the Palais–Smale condition due to the criticality of the embedding of $W^{2,2}(\mathbb{T}^4)$ into $W^{1,4}(\mathbb{T}^4)$. Even for this simple example, the existence theory of non-trivial critical points is completely open.

There are also fully non-linear versions of the prescribed scalar curvature problem; see e.g. [54]. These involve the *Schouten tensor*, defined as

$$A_g := \mathrm{Ric}_g - \frac{1}{2(n-1)} R_g g.$$

One is interested in prescribing as given functions the symmetric k-forms σ_k of the eigenvalues of A_g. For $k = 1, 2$ these are given by

$$\sigma_1 := \sum_{i=1}^{n} \lambda_i = \frac{n-2}{2(n-1)} R_g, \quad \sigma_2 := \lambda_1\lambda_2 + \lambda_1\lambda_3 + \cdots + \lambda_{n-1}\lambda_n,$$

so such prescription problems lead to extensions of (E_K).

Setting $\tilde{g} = e^{2u} g$, then $\sigma_k \equiv \mu$ if u solves

$$\sigma_k(A_g - \nabla^2 u + du \otimes du - \tfrac{1}{2}|\nabla u|^2 g) = \mu e^{2ku},$$

which is a PDE of fully non-linear character.

A metric g is said to *belong to the positive (respectively, negative) k-cone* if it satisfies $\sigma_j(A_g) > 0$ (respectively, $\sigma_j(-A_g) > 0$) for every $j = 1, 2, \dots, k$. A standard notation for this condition consists in writing that $g \in \Gamma_k^+$ (respectively, Γ_k^-), and under such circumstances the above equation becomes elliptic. When $k = 1, 2$, this equation is variational, but for $k \geq 3$ it is, in general, not variational (see [40, 215]). One of the interests for these problems is that if $g \in \Gamma_k^+$ (respectively, Γ_k^-) with $k > \frac{n}{2}$, then the Ricci tensor of g is positive definite (respectively, negative definite), which has relevant geometric consequences.

Constant σ_k-curvature has been achieved under the positive cone condition for $k = 2$ [56, 204], when (M, g) is locally conformally flat [118, 149], and for $k \geq \frac{n}{2}$ [123, 212]. The positive cone condition has been used to prove a priori estimates on solutions, and some counterexamples of local type or in domains were obtained in [155, 204] for the negative cone case. It is still possible that some global argument might rule out such counterexamples, but to the author's knowledge the existence problem in the negative cone is at the moment entirely open. Counterparts of the Kazdan–Warner problem in this context were studied e.g. in [58, 152–154].

10.4 Manifolds with boundary

A natural curvature prescription problem in the case of manifolds with boundary consists in also assigning the mean curvature of the boundary (the geodesic curvature, in two dimensions), when deforming the metric conformally.

If h_g stands for the mean curvature of the boundary with respect to the metric g, intended as the sum (not the average) of principal eigenvalues of the second fundamental form, or the geodesic curvature if $n = 2$, then one has the transformation laws

$$\frac{2}{n-2}\frac{\partial u}{\partial \nu} + h_g u = h_{\tilde{g}} u^{\frac{n}{n-2}} \quad \text{for } n \geq 3,$$
$$\frac{\partial u}{\partial \nu} + h_g = h_{\tilde{g}} e^u \quad \text{for } n = 2.$$

Here we have set, in our usual notation, $\tilde{g} = u^{\frac{4}{n-2}} g$ for $n \geq 3$ and $\tilde{g} = e^{2u}$ for $n = 2$, while ν stands for the outer unit normal at the boundary, taken with respect to g. Notice that the term at the boundary is also of critical type, since it involves the limiting trace embedding $H^1(M) \hookrightarrow L^{\frac{2(n-1)}{n-2}}(\partial M)$.

The problem was posed in [71, 72], where the regularity of weak solutions was proved, and some first existence results were obtained. The prescription of constant curvatures has been well studied, although not completely solved in general: we mention, among the literature, the papers [42, 69, 85, 98, 99, 125, 168, 175].

There are also several contributions about prescribing variable functions, at the interior or at the boundary, when one of the two is identically zero. We limit ourselves here to mentioning the papers [1, 2, 30, 94] (see also the references therein), which highlight some differences in the boundary blow-up behavior compared to the closed case: for example, imposing zero mean curvature on half three-dimensional spheres, multiple blow-up points might appear. In larger dimensions, even under assumptions analogous to those in Theorem 7.11, non-isolated blow-ups may occur. A quick explanation of this fact consists in the lack of the cancellation property (normally to the boundary) mentioned at the end of the proof of Proposition 6.10, which generates a stronger interaction of the bubbles with the prescribed curvature function compared to the closed case.

The problem with two variable functions has been less studied, also due to fact that, in dimension greater than or equal to 3, the Euler–Lagrange functional has a triple homogeneity, and hence it is not possible to employ weighted quotients as in (4.3). Some perturbative theorems are available in [6], while in [68] and [72] existence has been obtained up to some undetermined Lagrange multipliers. On the two-dimensional disk the recent paper [191] addresses the case of non-negative curvatures.

An interesting case, considered recently in [79, 80, 161], is when one prescribes negative scalar (or Gaussian) curvature in M and positive mean curvature on ∂M. If the prescribed scalar and mean curvature functions are $K(x) < 0$ at the interior and $H(x)$ at the boundary, then the function $\mathfrak{D}_n\colon \partial M \to \mathbb{R}$ defined as

$$\mathfrak{D}_n(x) := \sqrt{n(n-1)}\frac{H(x)}{\sqrt{|K(x)|}}, \quad x \in \partial M,$$

plays a crucial role in the study. In fact, blow-up at the boundary can only occur at points where $\mathfrak{D}_n \geq 1$, where the model geometry after blow-down is that of geodesic balls or *horospheres* in hyperbolic space; see Figure 10.3.

In the latter case the blow-up profile has infinite volume, and new techniques to deal with it had to be introduced. In two dimensions, there are indeed other blow-up profiles with infinite volume, classified in [109], which can be tackled using complex-analytic tools. Many questions still remain open, especially in dimension greater than 3. Some blow-up analysis results with upper bounds on the volume can be found in [28, 136]; it would also be interesting to construct blowing-up solutions with diverging volumes.

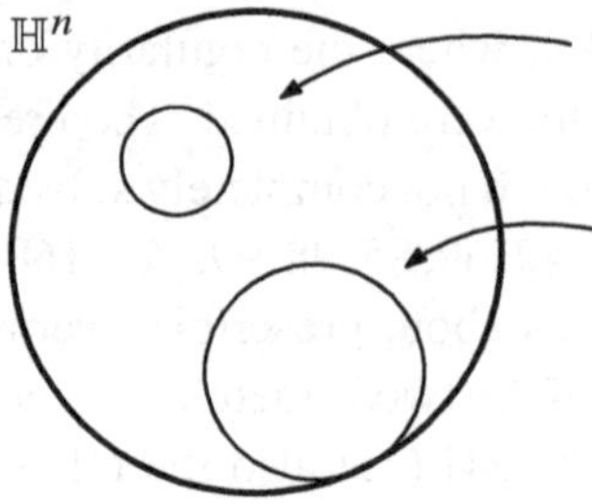

Figure 10.3. Bubbling profiles in negative curvature

10.5 The complete case

The conformal scalar curvature problem can of course also be posed on complete manifolds, but it is in general more difficult than the compact case due to lack of compactness for Sobolev's embeddings. This results in a more complicated phenomenology, and indeed the contributions in the literature are fewer and more sparse. For example, the Yamabe equation is sometimes not solvable [139]; see also [16, Chapter 5] for some existence results, especially concerning negative curvature. We also refer the reader to the recent book [170], where some existence, non-existence, and uniqueness results are presented. These typically involve assumptions on the curvatures and the volume growth of geodesic balls or spheres. Non-existence and uniqueness are established via a priori bounds on solutions, integral estimates proved via divergence and co-area formulas, combined with refined versions of the maximum principle; see e.g. [37] and [188]. Of particular relevance is the search of conformal metrics with given scalar curvature that preserve completeness of the metric. Existence is often derived via the sub-super-solution method, in the spirit of the classical Perron method for the Laplacian. Non-linear iteration schemes are employed, sometimes considering the equation on a sequence of compact sets that invade the whole manifold, and proving convergence of the corresponding solutions.

An important class of complete manifolds are the asymptotically flat ones; see e.g. [147], according to the following definition.

Definition 10.3. A Riemannian manifold (M^n, g) with $n \geq 3$ is said to be *asymptotically flat of order* $\tau > 0$ if there exists a compact set K such that $M \setminus K$ is an *end*, i.e. there exists a chart $\phi: M \setminus K \to \mathbb{R}^n \setminus B_1(0)$ such that the coordinate functions $(x^1, \ldots, x^n)$ with respect to this chart (called *asymptotic coordinates*) verify for the metric coefficients, as $\to +\infty$,

$$g_{ij}(x) = \delta_{ij} + O(|x|^{-\tau}), \quad \partial_k g_{ij}(x) = O(|x|^{-\tau-1}), \quad \partial^2_{kl} g_{ij}(x) = O(|x|^{-\tau-2}).$$

A similar definition can be given for asymptotically flat manifolds with multiple ends. An important example is *Schwarzschild's metric*, which has two ends and describes the gravitational field generated by a static black hole.

Associated to the above concept, there is that of *ADM mass*, introduced in [12],

$$m(g) := \frac{1}{2(n-1)|\mathbb{S}^{n-1}|} \lim_{\rho \to +\infty} \int_{S_\rho} \frac{x^j}{\rho} (\partial_i g_{ij} - \partial_j g_{ii}) \, d\sigma,$$

where S_ρ stands for the sphere $|x| = \rho$. It was proved in [27] that $m(g)$ is well defined, i.e. independent of the choice of a set of asymptotic coordinates as above provided that the order of flatness is sufficiently large, precisely if $\tau > \frac{n-2}{2}$. We also recall the positive mass theorem from [198, 199] (see also [217] for the case of spin manifolds), asserting that if $R_g \in L^1(M)$, $R_g \geq 0$, then $m(g) \geq 0$, with equality occurring only if (M, g) is isometric to the flat Euclidean space. We saw in Chapters 3 and 6 the relation of $m(g)$ to the Green's function of the conformal Laplacian L_g, and its fundamental role in preventing blow-up of solutions of critical equations or of minimizing sequences.

More precise results on the conformal prescription of scalar curvature are available for asymptotically flat manifolds; see e.g. [48, 171]. The Yamabe class for these manifolds can be defined in a similar manner as for the compact case, and can be shown to coincide with the one obtained by conformally compactifying the metric, similarly to what happens when conformally deforming $\mathbb{R}^n$ to $\mathbb{S}^n$ via the stereographic projection. Existence of conformal asymptotically flat metrics with constant scalar curvature (necessarily, zero) can only be obtained in positive Yamabe class, and in [87] some necessary and sufficient conditions are given in non-positive Yamabe class for the conformal prescription of a non-positive assigned scalar curvature function. There is also literature on other asymptotic geometries, but these will not be discussed here for reasons of brevity.

10.6 The Einstein constraint equations

The study of Yamabe-type equations also has interest in general relativity due to their role in *constraint equations*; see [51, 74]. Consider a Lorentzian manifold (N, h) of dimension $n + 1$, with metric h of signature $(-1, 1, \ldots, 1)$ and satisfying *Einstein's equation*

$$\mathrm{Ric}_h - \frac{1}{2} R_h h + \Lambda h = T.$$

Here, Ric_h and R_h stand for the Ricci tensor and the scalar curvature of h respectively, $\Lambda \in \mathbb{R}$ is the *cosmological constant*, while T, up to a physical constant, represents the stress-energy tensor.

Consider next a space-like hypersurface M embedded in N, let g be the Riemannian metric on M given by the pull-back of h, and let k be the second fundamental form of the immersion. The Gauss–Codazzi equations are given by the system

$$\begin{cases} R_g - \|k\|_g^2 + (\mathrm{tr}_g k)^2 = 2\mu + 2\Lambda, \\ \mathrm{div}_g(k - (\mathrm{tr}_g k)g) = J, \end{cases} \tag{10.1}$$

where μ and J are respectively a scalar quantity and a one-form determined by the tensor field T. Notice than when $k \equiv 0$ the first component simply reduces to the Yamabe equation.

As for the fundamental theorem of hypersurfaces in the Riemannian setting (see e.g. [205]), once a metric g and a symmetric 2-tensor k satisfying the above system are found in the vacuum case, it is possible to construct a local isometric immersion of (M, g) into (N, h) with second fundamental form equal to k, and such that the immersion induces the metric g on M by pull-back. The physical interest in such a statement is that, given a Riemannian manifold with data satisfying the above constraints, it is possible to locally extend it, uniquely within a proper class of solutions, as a Lorentzian spacetime via a system of hyperbolic equations; see [75, 107]. While the above system is in general undetermined, it is possible to produce solutions of special type.

One way to do it is to use *gluing methods*, which have counterparts in many problems arising in geometric analysis, like the construction of minimal or constant mean curvature surfaces and self-similar solutions to geometric evolution equations: we have already seen some applications in Chapter 7. Concerning the constraint equations, in [77] time-symmetric asymptotically flat metrics data of positive mass were glued to an element of the Schwarzschild family (handling the co-kernel via a Lyapunov–Schmidt reduction), with interesting extensions in e.g. [76] and [78]. A remarkable feature of this construction is the fact that the glued solution coincides with the original one on proper open sets. Other gluing constructions of initial data sets with this feature were performed in [52], where localized solutions with positive mass but exactly trivial outside suitable cones, exhibit a gravitational shielding phenomenon.

Another well-known strategy to construct solutions is to make instead a suitable ansatz, employing the so-called *conformal method*. Given a transverse (divergence-free) and trace-less symmetric two-covariant tensor σ and a scalar function γ, one looks for solutions of the above system of the following special form (displayed here in three dimensions):

$$\begin{cases} \bar{g} = u^4 g, \\ \bar{k} = u^{-2}(\sigma + \mathcal{K}_g W) + \frac{\gamma}{3} u^4 g. \end{cases}$$

In these formulas u is a positive function and $\mathcal{K}_g$ is the *conformal Killing operator* given by

$$\mathcal{K}_g W = \mathcal{L}_W g - \frac{1}{3}\mathrm{tr}_g(\mathcal{L}_W g) g,$$

where $\mathcal{L}_W$ stands for the Lie derivative. With this choice of unknowns u and W, for the special case $T = 0$ and $\Lambda = 0$, (10.1) reduces to

$$\begin{cases} L_g u = \frac{1}{8}\|\sigma + \mathcal{K}_g W\|_g^2 u^{-7} - \frac{1}{12}\gamma^2 u^5, \\ \mathrm{div}_g(\mathcal{K}_g W) = \frac{2}{3} u^6 d\gamma. \end{cases}$$

While there are several contributions to the study of the latter system, mostly using sub-super-solution methods, ODE analysis, and bifurcation theory, a complete understanding is only available for the *constant mean curvature* case, i.e. when γ is constant and W can be taken identically zero; see [135]. Several questions remain open, and we refer the reader to the recent survey [51] for an exhaustive and detailed description of the state of the art on the subject.

Appendix

Some useful results

In this appendix we collect some useful results, whose proofs involve non-trivial integration by parts. The first is Pohozaev's identity, useful to perform blow-up analysis. We then present the Kazdan–Warner identity, its spherical counterpart. We finally state and prove a rigidity result for Yamabe-type equations on Einstein's manifolds that relies on Bochner's identity.

A.1 Pohozaev's identity

A very useful tool to understand blow-ups of solutions is the following result:

Theorem A.1 (Pohozaev's identity). *Let $\Omega \subseteq \mathbb{R}^n$ be smooth and bounded, let $f(x,u)$ be a continuous function, and let $u \in C^2(\bar{\Omega})$ solve*

$$-\Delta u = f(x,u) \quad on\ \Omega.$$

Set

$$F(x,u) := \int_0^u f(x,t)\,dt.$$

Then

$$\int_\Omega \Big(nF(x,u) - \frac{n-2}{2} u f(x,u) + \sum_{i=1}^n x^i F_{x^i}(x,u) \Big)\,dx$$
$$= \int_{\partial\Omega} \Big(\langle x, \nu \rangle F(x,u) - \frac{1}{2}|\nabla u|^2 + \frac{\partial u}{\partial \nu} \langle x, \nu \rangle + \frac{n-2}{2} u \frac{\partial u}{\partial \nu} \Big)\,d\sigma,$$

where ν is the outer unit normal to $\partial\Omega$.

Proof. The proof consists in multiplying the equation by the scalar product $\langle x, \nabla u \rangle$ and integrating by parts. Using the notation $v_i := \partial_{x^i} v$, $v_{ij} := \partial^2_{x^i x^j} v$, and the convention of summing over repeated indices, we have

$$\int_\Omega \langle x, \nabla u \rangle f(x,u)\,dx$$
$$= \int_\Omega \langle x, \nabla F(x,u) \rangle\,dx - \int_\Omega x^i F_{x^i}(x,u)\,dx$$

$$= \int_\Omega \big[\operatorname{div}(xF(x,u)) - nF(x,u)\big]\,dx - \int_\Omega x^i F_{x^i}(x,u)\,dx$$
$$= \int_{\partial\Omega} F(x,u)\langle x,\nu\rangle\,d\sigma - \int_\Omega nF(x,u)\,dx - \int_\Omega x^i F_{x^i}(x,u)\,dx.$$

Moreover,

$$-\int_\Omega \Delta u\langle x,\nabla u\rangle\,dx = -\int_\Omega u_{jj}x^i u_i\,dx$$
$$= \int_\Omega u_j(x^i u_i)_j\,dx - \int_{\partial\Omega} u_j\nu_j x^i u_i\,d\sigma$$
$$= \int_\Omega (\delta_{ij}u_j u_i + x^i u_j u_{ij})\,dx - \int_{\partial\Omega}\langle\nabla u,\nu\rangle\langle x,\nabla u\rangle\,d\sigma$$
$$= \int_\Omega \Big(|\nabla u|^2 + \frac{1}{2}x^i(|\nabla u|^2)_i\Big)\,dx - \int_{\partial\Omega}\langle\nabla u,\nu\rangle\langle x,\nabla u\rangle\,d\sigma$$
$$= \frac{2-n}{2}\int_\Omega |\nabla u|^2\,dx + \frac{1}{2}\int_{\partial\Omega}|\nabla u|^2\langle x,\nu\rangle\,d\sigma - \int_{\partial\Omega}\langle\nabla u,\nu\rangle\langle x,\nabla u\rangle\,d\sigma.$$

We also have

$$\int_\Omega uf(x,u)\,dx = -\int_\Omega u\Delta u\,dx = -\int_\Omega (\operatorname{div}(u\nabla u) - |\nabla u|^2)\,dx$$
$$= \int_\Omega |\nabla u|^2\,dx - \int_{\partial\Omega} u\langle\nabla u,\nu\rangle\,d\sigma.$$

Combining the above identities, we obtain the desired conclusion. ■

We next specialize the above result to right-hand sides that are of power type.

Corollary A.2. *Suppose we are in the above situation with $\Omega = B_r$. Let $p \geq 1$ and let $u > 0$ be a C^2-solution of*

$$-\Delta u = \frac{1}{c_n}Ku^p \quad \text{on } B_r.$$

Then

$$\frac{1}{c_n(p+1)}\int_{B_r}\langle x,\nabla K\rangle u^{p+1}\,dx + \Big(\frac{n}{p+1} - \frac{n-2}{2}\Big)\frac{1}{c_n}\int_{B_r} Ku^{p+1}\,dx$$
$$- \frac{r}{c_n(p+1)}\int_{\partial B_r} Ku^{p+1}\,d\sigma$$
$$= \int_{\partial B_r}\mathcal{B}(r,x,u,\nabla u)\,d\sigma,$$

where

$$\mathcal{B}(r,x,u,\nabla u) = \frac{n-2}{2}u\frac{\partial u}{\partial\nu} - \frac{r}{2}|\nabla u|^2 + r\Big(\frac{\partial u}{\partial\nu}\Big)^2. \qquad \text{(A.1)}$$

The heuristics of the above proof rely on testing the equation $-\Delta u = f(x,u)$ on $\frac{d}{d\lambda}u(\lambda x)|_{\lambda=1} = \langle x, \nabla u\rangle$, to understand how the Euler–Lagrange functional varies when dilating the argument of a solution. We saw in Chapter 6 that highly concentrated solutions resemble small multiples of Green's functions away from the blow-up point. It is convenient next to evaluate $\mathcal{B}$ on such objects.

Proposition A.3. *We have the following properties:*

(i) *If* $u(x) = |x|^{2-n}$, *then* $\mathcal{B}(r,x,u,\nabla u) = 0$ *on* ∂B_r.

(ii) *If* $u(x) = |x|^{2-n} + A + \alpha(x)$ *for some* α *of class* C^1 *near* 0 *with* $\alpha(0) = 0$, *then there exists* $r^* > 0$ *such that for every* $r \in (0, r^*)$ *one has* $\mathcal{B}(r,x,u,\nabla u) < 0$ *on* ∂B_r. *Moreover,*

$$\lim_{r\to 0} \int_{\partial B_r} \mathcal{B}(r,x,u,\nabla u)\, d\sigma = -\frac{(n-2)^2}{2} A|\mathbb{S}^{n-1}|.$$

A.2 Kazdan–Warner identity

The next result is a *spherical version* of the previous one, allowing one to obtain forbidden functions for the prescribed scalar curvature problem.

Proposition A.4. *Let* $K \in C^1(\mathbb{S}^n)$, $n \geq 3$, *and let* u *be a positive solution to*

$$L_{g_{\mathbb{S}^n}} u = -c_n \Delta_{g_{\mathbb{S}^n}} u + R_{g_{\mathbb{S}^n}} u = K u^{\frac{n+2}{n-2}} \quad \text{on } \mathbb{S}^n.$$

Then

$$\int_{\mathbb{S}^n} g_{\mathbb{S}^n}(\nabla_{g_{\mathbb{S}^n}} K, \nabla_{g_{\mathbb{S}^n}} y^i) u^{\frac{2n}{n-2}}\, dV_{g_{\mathbb{S}^n}} = 0$$

for every $i = 1, \dots, n+1$, *where* $y^1, \dots, y^{n+1}$ *are the Euclidean coordinates of* $\mathbb{R}^{n+1}$ *restricted on* $\mathbb{S}^n$.

Proof. We first note the following identity for two smooth functions u and v on a Riemannian manifold (M, g):

$$\begin{aligned} 2\Delta_g u g(\nabla_g u, \nabla_g v) = {} & \operatorname{div}\big[2g(\nabla_g u, \nabla_g v)\nabla_g u - |\nabla_g u|_g^2 v\big] \\ & - (2\nabla_g^2 v - \Delta_g v g)(\nabla_g u, \nabla_g u), \end{aligned} \tag{A.2}$$

where ∇_g^2 denotes the Hessian. We recall that on the standard sphere $\mathbb{S}^n$, the function $\varphi := y^i$ satisfies

$$-\Delta_{g_{\mathbb{S}^n}} \varphi = n\varphi, \quad \nabla^2_{g_{\mathbb{S}^n}} \varphi = -\varphi g_{\mathbb{S}^n}.$$

Substituting this in (A.2) with $v = \varphi$, we get

$$2\Delta_{g_{\mathbb{S}^n}} u g_{\mathbb{S}^n}(\nabla_{g_{\mathbb{S}^n}} u, \nabla_{g_{\mathbb{S}^n}} \varphi) = \operatorname{div}_{g_{\mathbb{S}^n}} F - (n-2)\varphi|\nabla_{g_{\mathbb{S}^n}} u|^2_{g_{\mathbb{S}^n}},$$

where

$$F := 2g_{\mathbb{S}^n}(\nabla_{g_{\mathbb{S}^n}} u, \nabla_{g_{\mathbb{S}^n}} \varphi)\nabla_{g_{\mathbb{S}^n}} u - |\nabla_{g_{\mathbb{S}^n}} u|^2_{g_{\mathbb{S}^n}} \varphi.$$

Therefore,

$$2\int_{\mathbb{S}^n} \Delta_{g_{\mathbb{S}^n}} u g_{\mathbb{S}^n}(\nabla_{g_{\mathbb{S}^n}} u, \nabla_{g_{\mathbb{S}^n}} \varphi)\, dV_{g_{\mathbb{S}^n}} = -(n-2)\int_{\mathbb{S}^n} \varphi |\nabla_{g_{\mathbb{S}^n}} u|^2_{g_{\mathbb{S}^n}}\, dV_{g_{\mathbb{S}^n}}.$$

Moreover, using the identity

$$\frac{1}{2}\Delta_g u^2 = |\nabla_g u|^2_g + u\Delta_g u,$$

one finds

$$\begin{aligned}\int_{\mathbb{S}^n} \varphi u \Delta_{g_{\mathbb{S}^n}} u\, dV_{g_{\mathbb{S}^n}} &= -\int_{\mathbb{S}^n} \varphi |\nabla_{g_{\mathbb{S}^n}} u|^2_{g_{\mathbb{S}^n}}\, dV_{g_{\mathbb{S}^n}} + \frac{1}{2}\int_{\mathbb{S}^n} u^2 \Delta_{g_{\mathbb{S}^n}} \varphi\, dV_{g_{\mathbb{S}^n}} \\ &= -\int_{\mathbb{S}^n} \varphi |\nabla_{g_{\mathbb{S}^n}} u|^2_{g_{\mathbb{S}^n}}\, dV_{g_{\mathbb{S}^n}} - \frac{n}{2}\int_{\mathbb{S}^n} u^2 \varphi\, dV_{g_{\mathbb{S}^n}}.\end{aligned}$$

An integration by parts then leads to

$$\begin{aligned}&\int_{\mathbb{S}^n} g_{\mathbb{S}^n}(\nabla_{g_{\mathbb{S}^n}} K, \nabla_{g_{\mathbb{S}^n}} \varphi) u^{\frac{2n}{n-2}}\, dV_{g_{\mathbb{S}^n}} \\ &\quad = -\int_{\mathbb{S}^n} \Delta_{g_{\mathbb{S}^n}} \varphi (K u^{\frac{n+2}{n-2}}) u\, dV_{g_{\mathbb{S}^n}} \\ &\qquad - \frac{2n}{n-2}\int_{\mathbb{S}^n} g_{\mathbb{S}^n}(\nabla_{g_{\mathbb{S}^n}} \varphi, \nabla_{g_{\mathbb{S}^n}} u) K u^{\frac{n+2}{n-2}}\, dV_{g_{\mathbb{S}^n}} \\ &\quad = n\int_{\mathbb{S}^n} \varphi(-c_n u \Delta_{g_{\mathbb{S}^n}} u + R_{g_{\mathbb{S}^n}} u^2)\, dV_{g_{\mathbb{S}^n}} \\ &\qquad - \frac{2n}{n-2}\int_{\mathbb{S}^n} g_{\mathbb{S}^n}(\nabla_{g_{\mathbb{S}^n}} \varphi, \nabla_{g_{\mathbb{S}^n}} u)(-c_n \Delta_{g_{\mathbb{S}^n}} u + R_{g_{\mathbb{S}^n}} u)\, dV_{g_{\mathbb{S}^n}} \\ &\quad = -nc_n \int_{\mathbb{S}^n} \varphi u \Delta_{g_{\mathbb{S}^n}} u\, dV_{g_{\mathbb{S}^n}} + nR_{g_{\mathbb{S}^n}} \int_{\mathbb{S}^n} \varphi u^2\, dV_{g_{\mathbb{S}^n}} \\ &\qquad + \frac{2nc_n}{n-2}\int_{\mathbb{S}^n} g_{\mathbb{S}^n}(\nabla_{g_{\mathbb{S}^n}} \varphi, \nabla_{g_{\mathbb{S}^n}} u)\Delta_{g_{\mathbb{S}^n}} u\, dV_{g_{\mathbb{S}^n}} \\ &\qquad - \frac{nR_{g_{\mathbb{S}^n}}}{n-2}\int_{\mathbb{S}^n} g_{\mathbb{S}^n}(\nabla_{g_{\mathbb{S}^n}} \varphi, \nabla_{g_{\mathbb{S}^n}} u^2)\, dV_{g_{\mathbb{S}^n}} \\ &\quad = nc_n \int_{\mathbb{S}^n} \varphi |\nabla_{g_{\mathbb{S}^n}} u|^2_{g_{\mathbb{S}^n}}\, dV_{g_{\mathbb{S}^n}} + \Big(nR_{g_{\mathbb{S}^n}} + \frac{n^2 c_n}{2}\Big)\int_{\mathbb{S}^n} u^2 \varphi\, dV_{g_{\mathbb{S}^n}} \\ &\qquad - nc_n \int_{\mathbb{S}^n} \varphi |\nabla_{g_{\mathbb{S}^n}} u|^2_{g_{\mathbb{S}^n}}\, dV_{g_{\mathbb{S}^n}} + \frac{nR_{g_{\mathbb{S}^n}}}{n-2}\int_{\mathbb{S}^n} u^2 \Delta_{g_{\mathbb{S}^n}} \varphi\, dV_{g_{\mathbb{S}^n}} \\ &\quad = \Big(nR_{g_{\mathbb{S}^n}} + \frac{n^2 c_n}{2} - \frac{n^2 R_{g_{\mathbb{S}^n}}}{n-2}\Big)\int_{\mathbb{S}^n} u^2 \varphi\, dV_{g_{\mathbb{S}^n}} = 0,\end{aligned}$$

since $R_{g_{\mathbb{S}^n}} = n(n-1)$ and $c_n = 4\frac{n-1}{n-2}$. ∎

A.3 A rigidity result

We describe next a theorem implying constancy of solutions to sub-critical equations in Einstein manifolds, related to a classical one from [181].

Theorem A.5 ([34]). *Let (M, g) be a smooth Riemannian manifold of dimension $n \geq 2$. Let u be a positive solution of*

$$-\Delta_g u + \lambda u = u^q \quad \text{on } M,$$

for some $q > 1$ and $\lambda > 0$. Assume that

$$\mathrm{Ric}_g \geq \frac{n-1}{n}(q-1)\lambda g \quad \text{and} \quad q \leq \frac{n+2}{n-2}.$$

Suppose also that one of the above two inequalities is strict if $(M, g) = (\mathbb{S}^n, g_{\mathbb{S}^n})$. Then u is identically constant.

Proof. For $\beta \in \mathbb{R} \setminus \{0\}$, which will be chosen later, we set $u = v^{-\beta}$. Then v satisfies

$$\Delta_g v = (\beta + 1)\frac{|\nabla_g v|^2}{v} + \frac{1}{\beta}(v^{1+\beta-\beta q} - \lambda v). \tag{A.3}$$

Let us recall Bochner's formula:

$$\frac{1}{2}\Delta_g(|\nabla_g f|_g^2) = |\nabla_g^2 f|_g^2 + g(\nabla_g \Delta_g f, \nabla_g f) + \mathrm{Ric}_g(\nabla_g f, \nabla_g f). \tag{A.4}$$

Consider the above identity (A.4) for $f = v$, multiply it by v^γ for some $\gamma \in \mathbb{R}$ (which will also be chosen later) and integrate on M to find

$$\begin{aligned} 0 = &\int_M v^\gamma |\nabla_g^2 v|^2 \, dV_g + \int_M v^\gamma g(\nabla_g \Delta_g v, \nabla_g v) \, dV_g - \frac{1}{2}\int_M v^\gamma \Delta_g |\nabla_g v|^2 \, dV_g \\ &+ \int_M v^\gamma \mathrm{Ric}_g(\nabla_g v, \nabla_g v) \, dV_g. \end{aligned}$$

Using (A.3) we compute

$$\begin{aligned} &\int_M v^\gamma (\Delta_g v)^2 \, dV_g \\ &\quad = \int_M v^\gamma \Delta_g v \Big((\beta + 1)\frac{|\nabla_g v|^2}{v} + \frac{1}{c_n \beta}(v^{1+\beta-\beta q} - R_g v) \Big) \, dV_g \\ &\quad = (\beta + 1)\int_M v^{\gamma-1} |\nabla_g v|_g^2 \Big((\beta + 1)\frac{|\nabla_g v|^2}{v} + \frac{1}{\beta c_n}(v^{1+\beta-\beta q} - R_g v) \Big) \, dV_g \\ &\quad\quad + \frac{1}{c_n \beta}\int_M (v^{\gamma+1-\beta q+\beta} - R_g v^{\gamma+1}) \Delta_g v \, dV_g \end{aligned}$$

$$= (\beta+1)^1 \int_M v^{\gamma-2}|\nabla_g v|^4\,dV_g + \frac{\beta q-\gamma}{\beta}\int_M v^{\gamma+1-\beta q}|\nabla_g v|^2\,dV_g$$
$$+ \frac{\lambda(\gamma-\beta)}{\beta}\int_M v^\gamma|\nabla v|^2\,dV_g,$$

where the last equality follows from an integration by parts. We next compute

$$\int_M v^\gamma g(\nabla\Delta_g v, \nabla_g v)\,dV_g$$
$$= (\beta+1)\int_M g\Big(\nabla_g\frac{|\nabla_g v|^2}{v}, \nabla_g v\Big)\,dV_g$$
$$+ \frac{1+\beta-q\beta}{\beta}\int_M v^{\gamma+\beta-q\beta}|\nabla_g v|^2\,dV_g - \frac{\lambda}{\beta}\int_M v^\gamma|\nabla_g v|^2\,dV_g$$
$$= -(\beta+1)(\gamma+\beta+1)\int_M v^{\gamma-2}|\nabla v|^4\,dV_g$$
$$- q\int_M v^{\gamma+1-q\beta}|\nabla_g v|^2\,dV_g + \lambda\int_M v^\gamma|\nabla_g v|^2\,dV_g.$$

We also have

$$\int_M v^\gamma\Delta_g(|\nabla_g v|^2)\,dV_g$$
$$= \int_M |\nabla_g v|^2\Delta_g(v^\gamma)\,dV_g$$
$$= \int_M |\nabla v|^2(\gamma v^{\gamma-1}\Delta_g v + \gamma(\gamma-1)v^{\gamma-2}|\nabla_g v|^2)\,dV_g$$
$$= \gamma\int_M v^{\gamma-1}|\nabla_g v|^2\Big((\beta+1)\frac{|\nabla_g v|^2}{v} + \frac{1}{\beta}(v^{1+\beta-q\beta} - \lambda v)\Big)\,dV_g$$
$$+ \gamma(\gamma-1)\int_M v^{\gamma-2}|\nabla v|^4\,dV_g$$
$$= \gamma(\gamma+\beta)\int_M v^{\gamma-2}|\nabla_g v|^4\,dV_g + \frac{\gamma}{\beta}\int_M v^{\gamma+\beta-q\beta}|\nabla_g v|^2\,dV_g$$
$$- \frac{\gamma}{\beta}\int_M v^\gamma|\nabla v|^2\,dV_g.$$

Combining the above relations we obtain

$$\int_M v^\gamma\Big(|\nabla_g^2 v|^2 - \frac{1}{n}(\Delta_g v)^2\Big)\,dV_g$$
$$= a_1\int_M v^{\gamma-2}|\nabla v|^4\,dV_g + a_2\int_M v^{\gamma+\beta-q\beta}|\nabla_g v|^2\,dV_g$$
$$- \int_M v^\gamma(\mathrm{Ric}(\nabla_g v, \nabla_g v) + a_3|\nabla v|^2)\,dV_g, \tag{A.5}$$

where

$$a_1 := \frac{1}{2}\Big(\gamma^2 + (3\beta + 2)\gamma + 2\frac{n-1}{n}(\beta + 1)^2\Big),$$
$$a_2 := \frac{1}{n}\Big(q(n-1) + \frac{\gamma}{\beta}\frac{n+2}{2}\Big),$$
$$a_3 := \lambda\Big(\frac{n-1}{n} + \frac{\gamma}{\beta}\frac{n+2}{2n}\Big).$$

Since $|\nabla_g^2 v|^2 - \frac{1}{n}(\Delta_g v)^2 \geq 0$, the proof would follow if we could find $\beta \neq 0$ and $\gamma \in \mathbb{R}$ such that

$$a_1 \leq 0, \quad a_2 \leq 0, \quad \mathrm{Ric}_g + a_3 g \geq 0,$$

and at least one of the inequalities is strict. We set

$$b_1 := 1 + \frac{1}{\beta} \neq 1, \quad b_2 := -\frac{\gamma}{\beta} \in \mathbb{R}.$$

Then the above inequalities reduce to

$$2\frac{n-1}{n}b_1^2 - 2b_1b_2 + b_2^2 - b_2 \leq 0, \quad 2q\frac{n-1}{n+2} \leq b_2, \tag{A.6}$$
$$\frac{2n}{n+2}\,\mathrm{Ric}_g \geq \lambda\Big(b_2 - 2\frac{n-1}{n+2}\Big)g,$$

with at least one strict inequality. From the hypothesis on the lower bound of the Ricci curvature we have

$$\frac{2n}{n+2}\,\mathrm{Ric}_g \geq \lambda\Big(2q\frac{n-1}{n+2} - 2\frac{n-1}{n+2}\Big)g. \tag{A.7}$$

Notice that, in order to have a solution b_1 satisfying the first inequality in (A.6), we necessarily need the discriminant to be non-negative, that is

$$\frac{b_2}{n}(2(n-1) - (n-2)b_2) \geq 0. \tag{A.8}$$

For $n = 2$, the above inequality is strict, and therefore we can take $b_2 = 2q\frac{n-1}{n-2}$, and one can also find b_1 satisfying strict inequality in (A.6).

For $n \geq 3$ we consider the following cases:

Case 1 ($q < \frac{n-2}{n-2}$). We take $b_2 = 2q\frac{n-1}{n+2}$. Notice that inequality (A.8) is strict, and hence we can find b_1 satisfying strict inequality in (A.6).

Case 2 ($q = \frac{n-2}{n+2}$ and $(M, g) = (\mathbb{S}^n, g_{\mathbb{S}^n})$). In this case the inequality in (A.7) is strict, and we can take $b_2 = 2\frac{n-1}{n-2}$, $b_1 = \frac{n}{n-2}$.

Case 3 ($q = \frac{n+2}{n-2}$ and $(M, g) \neq (\mathbb{S}^n, g_{\mathbb{S}^n})$ and (A.7) is not strict). There exists $x_0 \in M$ and $e \in T_{x_0}M \setminus \{0\}$ such that

$$\mathrm{Ric}_g(e, e) = \frac{n-1}{n}(q-1)\lambda|e|^2,$$

and therefore for every non-constant function v,

$$\int_M v^\gamma \left(|\nabla_g^2 v|^2 - \frac{1}{n}(\Delta_g v)^2 \right) dV_g > 0.$$

Taking $a_1 = a_2 = 0$, we see that the left-hand side of (A.5) is positive unless v is identically constant. ■

References

[1] W. Abdelhedi, H. Chtioui, and M. O. Ahmedou, A Morse theoretical approach for the boundary mean curvature problem on $\mathbb{B}^4$. *J. Funct. Anal.* **254** (2008), no. 5, 1307–1341

[2] M. Ahmedou and M. Ben Ayed, Non simple blow ups for the Nirenberg problem on half spheres. *Discrete Contin. Dyn. Syst.* **42** (2022), no. 12, 5967–6005

[3] R. Albesiano, Solutions of Liouville equations with non-trivial profile in dimensions 2 and 4. *J. Differential Equations* **272** (2021), 606–647

[4] A. D. Aleksandrov, Uniqueness theorems for surfaces in the large. V. *Vestnik Leningrad. Univ.* **13** (1958), no. 19, 5–8

[5] A. Ambrosetti, J. Garcia Azorero, and I. Peral, Perturbation of $\Delta u + u^{(N+2)/(N-2)} = 0$, the scalar curvature problem in $\mathbf{R}^N$, and related topics. *J. Funct. Anal.* **165** (1999), no. 1, 117–149

[6] A. Ambrosetti, Y. Li, and A. Malchiodi, On the Yamabe problem and the scalar curvature problems under boundary conditions. *Math. Ann.* **322** (2002), no. 4, 667–699

[7] A. Ambrosetti and A. Malchiodi, A multiplicity result for the Yamabe problem on S^n. *J. Funct. Anal.* **168** (1999), no. 2, 529–561

[8] A. Ambrosetti and A. Malchiodi, On the symmetric scalar curvature problem on S^n. *J. Differential Equations* **170** (2001), no. 1, 228–245

[9] A. Ambrosetti and A. Malchiodi, *Perturbation methods and semilinear elliptic problems on* $\mathbf{R}^n$. Prog. Math. 240, Birkhäuser, Basel, 2006

[10] A. Ambrosetti and A. Malchiodi, *Nonlinear analysis and semilinear elliptic problems*. Camb. Stud. Adv. Math. 104, Cambridge University Press, Cambridge, 2007

[11] A. Ambrosetti and G. Prodi, *A primer of nonlinear analysis*. Camb. Stud. Adv. Math. 34, Cambridge University Press, Cambridge, 1993

[12] R. Arnowitt, S. Deser, and C. W. Misner, Energy and the criteria for radiation in general relativity. *Phys. Rev. (2)* **118** (1960), 1100–1104

[13] T. Aubin, Métriques riemanniennes et courbure. *J. Differential Geometry* **4** (1970), 383–424

[14] T. Aubin, Isoperimetric problems and Sobolev spaces. *J. Differential Geometry* **11** (1976), no. 4, 573–598

[15] T. Aubin, Nonlinear differential equations and Yamabe problem concerning scalar curvature. *J. Math. Pure Appl. (9)* **55** (1976), no. 3, 269–296

[16] T. Aubin, *Some nonlinear problems in Riemannian geometry*. Springer Monogr. Math., Springer, Berlin, 1998

[17] T. Aubin and A. Bahri, Méthodes de topologie algébrique pour le problème de la courbure scalaire prescrite. *J. Math. Pures Appl. (9)* **76** (1997), no. 6, 525–549

[18] T. Aubin and A. Bahri, Une hypothèse topologique pour le problème de la courbure scalaire prescrite. *J. Math. Pures Appl. (9)* **76** (1997), no. 10, 843–850

[19] T. Aubin and S. Bismuth, Courbure scalaire prescrite sur les variétés riemanniennes compactes dans le cas négatif. *J. Funct. Anal.* **143** (1997), no. 2, 529–541

[20] A. Bahri, *Critical points at infinity in some variational problems*. Pitman Res. Notes Math. Ser. 182, Longman Scientific & Technical, Harlow; copublished in the United States with John Wiley & Sons, New York, 1989

[21] A. Bahri, An invariant for Yamabe-type flows with applications to scalar-curvature problems in high dimension. *Duke Math. J.* **81** (1996), no. 2, 323–466

[22] A. Bahri and H. Brezis, Non-linear elliptic equations on Riemannian manifolds with the Sobolev critical exponent. In *Topics in geometry*, pp. 1–100, Progr. Nonlinear Differential Equations Appl. 20, Birkhäuser Boston, Boston, MA, 1996

[23] A. Bahri and J.-M. Coron, On a nonlinear elliptic equation involving the critical Sobolev exponent: The effect of the topology of the domain. *Comm. Pure Appl. Math.* **41** (1988), no. 3, 253–294

[24] A. Bahri and J.-M. Coron, The scalar-curvature problem on the standard three-dimensional sphere. *J. Funct. Anal.* **95** (1991), no. 1, 106–172

[25] A. Bahri, Y. Li, and O. Rey, On a variational problem with lack of compactness: The topological effect of the critical points at infinity. *Calc. Var. Partial Differential Equations* **3** (1995), no. 1, 67–93

[26] P. Baird, A. Fardoun, and R. Regbaoui, Q-curvature flow on 4-manifolds. *Calc. Var. Partial Differential Equations* **27** (2006), no. 1, 75–104

[27] R. Bartnik, The mass of an asymptotically flat manifold. *Comm. Pure Appl. Math.* **39** (1986), no. 5, 661–693

[28] L. Battaglia, M. Medina, and A. Pistoia, Large conformal metrics with prescribed Gaussian and geodesic curvatures. *Calc. Var. Partial Differential Equations* **60** (2021), no. 1, article no. 39

[29] M. Ben Ayed, Y. Chen, H. Chtioui, and M. Hammami, On the prescribed scalar curvature problem on 4-manifolds. *Duke Math. J.* **84** (1996), no. 3, 633–677

[30] M. Ben Ayed, K. El Mehdi, and M. Ould Ahmedou, The scalar curvature problem on the four dimensional half sphere. *Calc. Var. Partial Differential Equations* **22** (2005), no. 4, 465–482

[31] M. Berti and A. Malchiodi, Non-compactness and multiplicity results for the Yamabe problem on S^n. *J. Funct. Anal.* **180** (2001), no. 1, 210–241

[32] A. L. Besse, *Einstein manifolds*. Class. Math., Springer, Berlin, 2008

[33] G. Bianchi, The scalar curvature equation on $\mathbf{R}^n$ and S^n. *Adv. Differential Equations* **1** (1996), no. 5, 857–880

[34] M.-F. Bidaut-Véron and L. Véron, Nonlinear elliptic equations on compact Riemannian manifolds and asymptotics of Emden equations. *Invent. Math.* **106** (1991), no. 3, 489–539

[35] F. Borer, L. Galimberti, and M. Struwe, "Large" conformal metrics of prescribed Gauss curvature on surfaces of higher genus. *Comment. Math. Helv.* **90** (2015), no. 2, 407–428

[36] J.-P. Bourguignon and J.-P. Ezin, Scalar curvature functions in a conformal class of metrics and conformal transformations. *Trans. Amer. Math. Soc.* **301** (1987), no. 2, 723–736

[37] L. Brandolini, M. Rigoli, and A. G. Setti, Positive solutions of Yamabe type equations on complete manifolds and applications. *J. Funct. Anal.* **160** (1998), no. 1, 176–222

[38] T. P. Branson, Differential operators canonically associated to a conformal structure. *Math. Scand.* **57** (1985), no. 2, 293–345

[39] T. P. Branson, S.-Y. A. Chang, and P. C. Yang, Estimates and extremals for zeta function determinants on four-manifolds. *Comm. Math. Phys.* **149** (1992), no. 2, 241–262

[40] T. P. Branson and A. R. Gover, Variational status of a class of fully nonlinear curvature prescription problems. *Calc. Var. Partial Differential Equations* **32** (2008), no. 2, 253–262

[41] T. P. Branson and B. Ørsted, Explicit functional determinants in four dimensions. *Proc. Amer. Math. Soc.* **113** (1991), no. 3, 669–682

[42] S. Brendle, Curvature flows on surfaces with boundary. *Math. Ann.* **324** (2002), no. 3, 491–519

[43] S. Brendle and F. C. Marques, Blow-up phenomena for the Yamabe equation. II. *J. Differential Geom.* **81** (2009), no. 2, 225–250

[44] A. Bressan, *Hyperbolic systems of conservation laws*. Oxf. Lect. Ser. Math. Appl. 20, Oxford University Press, Oxford, 2000

[45] H. Brezis, *Functional analysis, Sobolev spaces and partial differential equations*. Universitext, Springer, New York, 2011

[46] H. Brézis and E. Lieb, A relation between pointwise convergence of functions and convergence of functionals. *Proc. Amer. Math. Soc.* **88** (1983), no. 3, 486–490

[47] L. A. Caffarelli, B. Gidas, and J. Spruck, Asymptotic symmetry and local behavior of semilinear elliptic equations with critical Sobolev growth. *Comm. Pure Appl. Math.* **42** (1989), no. 3, 271–297

[48] M. Cantor, The existence of non-trivial asymptotically flat initial data for vacuum spacetimes. *Comm. Math. Phys.* **57** (1977), no. 1, 83–96

[49] D. Cao, E. S. Noussair, and S. Yan, On the scalar curvature equation $-\Delta u = (1 + \varepsilon K)u^{(N+2)/(N-2)}$ in $\mathbb{R}^N$. *Calc. Var. Partial Differential Equations* **15** (2002), no. 3, 403–419

[50] A. Carlotto, A survey on positive scalar curvature metrics. *Boll. Unione Mat. Ital.* **14** (2021), no. 1, 17–42

[51] A. Carlotto, The general relativistic constraint equations. *Living Rev. Relativ.* **24** (2021), article no. 2

[52] A. Carlotto and R. Schoen, Localizing solutions of the Einstein constraint equations. *Invent. Math.* **205** (2016), no. 3, 559–615

[53] K.-c. Chang, *Infinite-dimensional Morse theory and multiple solution problems*. Prog. Nonlinear Differ. Equ. Appl. 6, Birkhäuser Boston, Boston, MA, 1993

[54] S.-Y. A. Chang, *Non-linear elliptic equations in conformal geometry*. Zur. Lect. Adv. Math., European Mathematical Society (EMS), Zürich, 2004

[55] S.-Y. A. Chang and M. d. M. González, Fractional Laplacian in conformal geometry. *Adv. Math.* **226** (2011), no. 2, 1410–1432

[56] S.-Y. A. Chang, M. J. Gursky, and P. Yang, An a priori estimate for a fully nonlinear equation on four-manifolds. *J. Anal. Math.* **87** (2002), 151–186

[57] S.-Y. A. Chang, M. J. Gursky, and P. C. Yang, The scalar curvature equation on 2- and 3-spheres. *Calc. Var. Partial Differential Equations* **1** (1993), no. 2, 205–229

[58] S.-Y. A. Chang, Z.-C. Han, and P. Yang, On the prescribing σ_2 curvature equation on $\mathbb{S}^4$. *Calc. Var. Partial Differential Equations* **40** (2011), no. 3-4, 539–565

[59] S.-Y. A. Chang and P. C. Yang, Prescribing Gaussian curvature on S^2. *Acta Math.* **159** (1987), no. 3-4, 215–259

[60] S.-Y. A. Chang and P. C. Yang, Conformal deformation of metrics on S^2. *J. Differential Geom.* **27** (1988), no. 2, 259–296

[61] S.-Y. A. Chang and P. C. Yang, Extremal metrics of zeta function determinants on 4-manifolds. *Ann. of Math. (2)* **142** (1995), no. 1, 171–212

[62] C.-C. Chen and C.-S. Lin, Estimates of the conformal scalar curvature equation via the method of moving planes. *Comm. Pure Appl. Math.* **50** (1997), no. 10, 971–1017

[63] C.-C. Chen and C.-S. Lin, Estimate of the conformal scalar curvature equation via the method of moving planes. II. *J. Differential Geom.* **49** (1998), no. 1, 115–178

[64] C.-C. Chen and C.-S. Lin, Blowing up with infinite energy of conformal metrics on S^n. *Comm. Partial Differential Equations* **24** (1999), no. 5-6, 785–799

[65] W. Chen and C. Li, *Methods on nonlinear elliptic equations*. AIMS Ser. Differ. Equ. Dyn. Syst. 4, American Institute of Mathematical Sciences (AIMS), Springfield, MO, 2010

[66] W. X. Chen and W. Y. Ding, Scalar curvatures on S^2. *Trans. Amer. Math. Soc.* **303** (1987), no. 1, 365–382

[67] W. X. Chen and C. Li, Classification of solutions of some nonlinear elliptic equations. *Duke Math. J.* **63** (1991), no. 3, 615–622

[68] X. Chen, P. T. Ho, and L. Sun, Prescribed scalar curvature plus mean curvature flows in compact manifolds with boundary of negative conformal invariant. *Ann. Global Anal. Geom.* **53** (2018), no. 1, 121–150

[69] X. Chen, Y. Ruan, and L. Sun, The Han–Li conjecture in constant scalar curvature and constant boundary mean curvature problem on compact manifolds. *Adv. Math.* **358** (2019), article no. 106854

[70] X. Chen and X. Xu, The scalar curvature flow on S^n—perturbation theorem revisited. *Invent. Math.* **187** (2012), no. 2, 395–506; Erratum to: The scalar curvature flow on S^n—perturbation theorem revisited. *Invent. Math.* **187** (2012), no. 2, 507–509

[71] P. Cherrier, Meilleures constantes dans des inégalités relatives aux espaces de Sobolev. *Bull. Sci. Math. (2)* **108** (1984), no. 3, 225–262

[72] P. Cherrier, Problèmes de Neumann non linéaires sur les variétés riemanniennes. *J. Funct. Anal.* **57** (1984), no. 2, 154–206

[73] M. Chipot, I. Shafrir, and M. Fila, On the solutions to some elliptic equations with nonlinear Neumann boundary conditions. *Adv. Differential Equations* **1** (1996), no. 1, 91–110

[74] Y. Choquet-Bruhat, *General relativity and the Einstein equations*. Oxford Math. Monogr., Oxford University Press, Oxford, 2009

[75] Y. Choquet-Bruhat and R. Geroch, Global aspects of the Cauchy problem in general relativity. *Comm. Math. Phys.* **14** (1969), 329–335

[76] P. T. Chruściel and E. Delay, On mapping properties of the general relativistic constraints operator in weighted function spaces, with applications. *Mém. Soc. Math. Fr. (N.S.)* (2003), no. 94

[77] J. Corvino, Scalar curvature deformation and a gluing construction for the Einstein constraint equations. *Comm. Math. Phys.* **214** (2000), no. 1, 137–189

[78] J. Corvino and R. M. Schoen, On the asymptotics for the vacuum Einstein constraint equations. *J. Differential Geom.* **73** (2006), no. 2, 185–217

[79] S. Cruz-Blázquez, A. Malchiodi, and D. Ruiz, Conformal metrics with prescribed scalar and mean curvature. *J. Reine Angew. Math.* **789** (2022), 211–251

[80] S. Cruz-Blázquez and D. Ruiz, Prescribing Gaussian and geodesic curvature on disks. *Adv. Nonlinear Stud.* **18** (2018), no. 3, 453–468

[81] K. Deimling, *Nonlinear functional analysis*. Springer, Berlin, 1985

[82] M. del Pino, P. Felmer, and M. Musso, Two-bubble solutions in the super-critical Bahri–Coron's problem. *Calc. Var. Partial Differential Equations* **16** (2003), no. 2, 113–145

[83] M. del Pino, M. Musso, F. Pacard, and A. Pistoia, Large energy entire solutions for the Yamabe equation. *J. Differential Equations* **251** (2011), no. 9, 2568–2597

[84] M. del Pino and C. Román, Large conformal metrics with prescribed sign-changing Gauss curvature. *Calc. Var. Partial Differential Equations* **54** (2015), no. 1, 763–789

[85] S. de Moura Almaraz, An existence theorem of conformal scalar-flat metrics on manifolds with boundary. *Pacific J. Math.* **248** (2010), no. 1, 1–22

[86] B. Deng, L. Sun, and J. Wei, Sharp quantitative estimates of Struwe's decomposition. 2021, arXiv:2103.15360

[87] J. Dilts and D. Maxwell, Yamabe classification and prescribed scalar curvature in the asymptotically Euclidean setting. *Comm. Anal. Geom.* **26** (2018), no. 5, 1127–1168

[88] W. Ding, A proof of the uniformization theorem on S^2. *J. Partial Differential Equations* **14** (2001), no. 3, 247–250

[89] W. Y. Ding, On a conformally invariant elliptic equation on $\mathbf{R}^n$. *Comm. Math. Phys.* **107** (1986), no. 2, 331–335

[90] Z. Djadli, E. Hebey, and M. Ledoux, Paneitz-type operators and applications. *Duke Math. J.* **104** (2000), no. 1, 129–169

[91] Z. Djadli and A. Malchiodi, Existence of conformal metrics with constant Q-curvature. *Ann. of Math. (2)* **168** (2008), no. 3, 813–858

[92] Z. Djadli, A. Malchiodi, and M. O. Ahmedou, Prescribing a fourth order conformal invariant on the standard sphere. I. A perturbation result. *Commun. Contemp. Math.* **4** (2002), no. 3, 375–408

[93] Z. Djadli, A. Malchiodi, and M. O. Ahmedou, Prescribing a fourth order conformal invariant on the standard sphere. II. Blow up analysis and applications. *Ann. Sc. Norm. Super. Pisa Cl. Sci. (5)* **1** (2002), no. 2, 387–434

[94] Z. Djadli, A. Malchiodi, and M. Ould Ahmedou, Prescribing scalar and boundary mean curvature on the three dimensional half sphere. *J. Geom. Anal.* **13** (2003), no. 2, 255–289

[95] M. P. do Carmo, *Differential geometry of curves & surfaces*. Dover Publications, Mineola, NY, 2016

[96] O. Druet, Compactness for Yamabe metrics in low dimensions. *Int. Math. Res. Not.* (2004), no. 23, 1143–1191

[97] O. Druet, E. Hebey, and F. Robert, *Blow-up theory for elliptic PDEs in Riemannian geometry*. Math. Notes (Princeton) 45, Princeton University Press, Princeton, NJ, 2004

[98] J. F. Escobar, Conformal deformation of a Riemannian metric to a scalar flat metric with constant mean curvature on the boundary. *Ann. of Math. (2)* **136** (1992), no. 1, 1–50

[99] J. F. Escobar, The Yamabe problem on manifolds with boundary. *J. Differential Geom.* **35** (1992), no. 1, 21–84

[100] J. F. Escobar and R. M. Schoen, Conformal metrics with prescribed scalar curvature. *Invent. Math.* **86** (1986), no. 2, 243–254

[101] P. Esposito and A. Malchiodi, Critical metrics for log-determinant functionals in conformal geometry. arXiv:1906.08188. To appear in *J. Differential Geom.*

[102] H. Federer, Some integralgeometric theorems. *Trans. Amer. Math. Soc.* **77** (1954), 238–261

[103] C. Fefferman and C. R. Graham, Q-curvature and Poincaré metrics. *Math. Res. Lett.* **9** (2002), no. 2-3, 139–151

[104] V. Felli, Existence of conformal metrics on S^n with prescribed fourth-order invariant. *Adv. Differential Equations* **7** (2002), no. 1, 47–76

[105] A. Figalli and F. Glaudo, On the sharp stability of critical points of the Sobolev inequality. *Arch. Ration. Mech. Anal.* **237** (2020), no. 1, 201–258

[106] A. Figalli and R. Neumayer, Gradient stability for the Sobolev inequality: The case $p \geq 2$. *J. Eur. Math. Soc. (JEMS)* **21** (2019), no. 2, 319–354

[107] Y. Fourès-Bruhat, Théorème d'existence pour certains systèmes d'équations aux dérivées partielles non linéaires. *Acta Math.* **88** (1952), 141–225

[108] R. H. Fowler, Further studies on Emden's and similar differential equations. *Q. J. Math.* **2** (1931), no. 1, 259–288

[109] J. A. Gálvez and P. Mira, The Liouville equation in a half-plane. *J. Differential Equations* **246** (2009), no. 11, 4173–4187

[110] L. Z. Gao and S.-T. Yau, The existence of negatively Ricci curved metrics on three-manifolds. *Invent. Math.* **85** (1986), no. 3, 637–652

[111] B. Gidas, W. M. Ni, and L. Nirenberg, Symmetry and related properties via the maximum principle. *Comm. Math. Phys.* **68** (1979), no. 3, 209–243

[112] B. Gidas and J. Spruck, Global and local behavior of positive solutions of nonlinear elliptic equations. *Comm. Pure Appl. Math.* **34** (1981), no. 4, 525–598

[113] D. Gilbarg and N. S. Trudinger, *Elliptic partial differential equations of second order*. Class. Math., Springer, Berlin, 2001

[114] C. R. Graham, R. Jenne, L. J. Mason, and G. A. J. Sparling, Conformally invariant powers of the Laplacian. I. Existence. *J. London Math. Soc. (2)* **46** (1992), no. 3, 557–565

[115] C. R. Graham and M. Zworski, Scattering matrix in conformal geometry. *Invent. Math.* **152** (2003), no. 1, 89–118

[116] M. Gromov and H. B. Lawson, Jr., The classification of simply connected manifolds of positive scalar curvature. *Ann. of Math. (2)* **111** (1980), no. 3, 423–434

[117] H.-C. Grunau and F. Robert, Positivity and almost positivity of biharmonic Green's functions under Dirichlet boundary conditions. *Arch. Ration. Mech. Anal.* **195** (2010), no. 3, 865–898

[118] P. Guan and G. Wang, Local estimates for a class of fully nonlinear equations arising from conformal geometry. *Int. Math. Res. Not.* (2003), no. 26, 1413–1432

[119] M. Gursky and A. Malchiodi, Non-uniqueness results for critical metrics of regularized determinants in four dimensions. *Comm. Math. Phys.* **315** (2012), no. 1, 1–37

[120] M. J. Gursky, Uniqueness of the functional determinant. *Comm. Math. Phys.* **189** (1997), no. 3, 655–665

[121] M. J. Gursky, The principal eigenvalue of a conformally invariant differential operator, with an application to semilinear elliptic PDE. *Comm. Math. Phys.* **207** (1999), no. 1, 131–143

[122] M. J. Gursky and A. Malchiodi, A strong maximum principle for the Paneitz operator and a non-local flow for the Q-curvature. *J. Eur. Math. Soc. (JEMS)* **17** (2015), no. 9, 2137–2173

[123] M. J. Gursky and J. A. Viaclovsky, A fully nonlinear equation on four-manifolds with positive scalar curvature. *J. Differential Geom.* **63** (2003), no. 1, 131–154

[124] Z.-C. Han, Prescribing Gaussian curvature on S^2. *Duke Math. J.* **61** (1990), no. 3, 679–703

[125] Z.-C. Han and Y. Li, The Yamabe problem on manifolds with boundary: Existence and compactness results. *Duke Math. J.* **99** (1999), no. 3, 489–542

[126] F. Hang and P. C. Yang, Q-curvature on a class of manifolds with dimension at least 5. *Comm. Pure Appl. Math.* **69** (2016), no. 8, 1452–1491

[127] A. Hatcher, *Algebraic topology*. Cambridge University Press, Cambridge, 2002

[128] E. Hebey, Changements de métriques conformes sur la sphère. Le problème de Nirenberg. *Bull. Sci. Math.* **114** (1990), no. 2, 215–242

[129] E. Hebey, *Sobolev spaces on Riemannian manifolds*. Lect. Notes Math. 1635, Springer, Berlin, 1996

[130] E. Hebey, *Nonlinear analysis on manifolds: Sobolev spaces and inequalities*. Courant Lect. Notes Math. 5, New York University, Courant Institute of Mathematical Sciences, New York; American Mathematical Society, Providence, RI, 1999

[131] E. Hebey, *Compactness and stability for nonlinear elliptic equations*. Zur. Lect. Adv. Math., European Mathematical Society (EMS), Zürich, 2014

[132] E. Hebey and M. Vaugon, Meilleures constantes dans le théorème d'inclusion de Sobolev et multiplicité pour les problèmes de Nirenberg et Yamabe. *Indiana Univ. Math. J.* **41** (1992), no. 2, 377–407

[133] E. Hebey and M. Vaugon, Le problème de Yamabe équivariant. *Bull. Sci. Math.* **117** (1993), no. 2, 241–286

[134] A. Hyder, S. Iula, and L. Martinazzi, Large blow-up sets for the prescribed Q-curvature equation in the Euclidean space. *Commun. Contemp. Math.* **20** (2018), no. 2, article no. 1750026

[135] J. Isenberg, Constant mean curvature solutions of the Einstein constraint equations on closed manifolds. *Classical Quantum Gravity* **12** (1995), no. 9, 2249–2274

[136] A. Jevnikar, R. López-Soriano, M. Medina, and D. Ruiz, Blow-up analysis of conformal metrics of the disk with prescribed Gaussian and geodesic curvatures. *Anal. PDE* **15** (2022), no. 8, 1897–1931

[137] T. Jin, Y. Li, and J. Xiong, On a fractional Nirenberg problem, part I: Blow up analysis and compactness of solutions. *J. Eur. Math. Soc. (JEMS)* **16** (2014), no. 6, 1111–1171

[138] T. Jin, Y. Y. Li, and J. Xiong, On a fractional Nirenberg problem, part II: Existence of solutions. *Int. Math. Res. Not. IMRN* (2015), no. 6, 1555–1589

[139] Z. R. Jin, A counterexample to the Yamabe problem for complete noncompact manifolds. In *Partial differential equations (Tianjin, 1986)*, pp. 93–101, Lect. Notes Math. 1306, Springer, Berlin, 1988

[140] J. L. Kazdan and F. W. Warner, Curvature functions for compact 2-manifolds. *Ann. of Math. (2)* **99** (1974), 14–47

[141] J. L. Kazdan and F. W. Warner, A direct approach to the determination of Gaussian and scalar curvature functions. *Invent. Math.* **28** (1975), 227–230

[142] J. L. Kazdan and F. W. Warner, Existence and conformal deformation of metrics with prescribed Gaussian and scalar curvatures. *Ann. of Math. (2)* **101** (1975), 317–331

[143] J. L. Kazdan and F. W. Warner, Prescribing curvatures. In *Differential geometry (Proc. Sympos. Pure Math., Vol. XXVII, Part 2, Stanford Univ., Stanford, Calif., 1973)*, pp. 309–319, American Mathematical Society, Providence, RI, 1975

[144] J. L. Kazdan and F. W. Warner, Scalar curvature and conformal deformation of Riemannian structure. *J. Differential Geometry* **10** (1975), 113–134

[145] M. A. Khuri, F. C. Marques, and R. M. Schoen, A compactness theorem for the Yamabe problem. *J. Differential Geom.* **81** (2009), no. 1, 143–196

[146] N. Korevaar, R. Mazzeo, F. Pacard, and R. Schoen, Refined asymptotics for constant scalar curvature metrics with isolated singularities. *Invent. Math.* **135** (1999), no. 2, 233–272

[147] J. M. Lee and T. H. Parker, The Yamabe problem. *Bull. Amer. Math. Soc. (N.S.)* **17** (1987), no. 1, 37–91

[148] M. C. Leung and F. Zhou, Conformal scalar curvature equation on S^n: Functions with two close critical points (twin pseudo-peaks). *Commun. Contemp. Math.* **20** (2018), no. 5, article no. 1750051

[149] A. Li and Y. Li, On some conformally invariant fully nonlinear equations. *Comm. Pure Appl. Math.* **56** (2003), no. 10, 1416–1464

[150] J. Li, Y. Li, and P. Liu, The Q-curvature on a 4-dimensional Riemannian manifold (M, g) with $\int_M Q dV_g = 8\pi^2$. *Adv. Math.* **231** (2012), no. 3-4, 2194–2223

[151] Y. Li, Prescribing scalar curvature on S^n and related problems. II. Existence and compactness. *Comm. Pure Appl. Math.* **49** (1996), no. 6, 541–597

[152] Y. Li, H. Lu, and S. Lu, On the σ_2-Nirenberg problem on $\mathbb{S}^2$. *J. Funct. Anal.* **283** (2022), no. 10, article no. 109606

[153] Y. Li, L. Nguyen, and B. Wang, The axisymmetric σ_k-Nirenberg problem. *J. Funct. Anal.* **281** (2021), no. 9, article no. 109198

[154] Y. Li, L. Nguyen, and B. Wang, On the σ_k-Nirenberg problem. arXiv:2008.08437. To appear in *Amer. J. Math.*

[155] Y. Li, L. Nguyen, and J. Xiong, Regularity of viscosity solutions of the σ_k-Loewner–Nirenberg problem. *Proc. London Math. Soc. (3)* **127** (2023), no. 1, 1–34

[156] Y. Li and L. Zhang, Liouville-type theorems and Harnack-type inequalities for semilinear elliptic equations. *J. Anal. Math.* **90** (2003), 27–87

[157] Y. Li and M. Zhu, Yamabe type equations on three-dimensional Riemannian manifolds. *Commun. Contemp. Math.* **1** (1999), no. 1, 1–50

[158] Y. Y. Li, Prescribing scalar curvature on S^n and related problems. I. *J. Differential Equations* **120** (1995), no. 2, 319–410

[159] Y. Y. Li and L. Zhang, Compactness of solutions to the Yamabe problem. II. *Calc. Var. Partial Differential Equations* **24** (2005), no. 2, 185–237

[160] J. Lohkamp, Negatively Ricci curved manifolds. *Bull. Amer. Math. Soc. (N.S.)* **27** (1992), no. 2, 288–291

[161] R. López-Soriano, A. Malchiodi, and D. Ruiz, Conformal metrics with prescribed Gaussian and geodesic curvatures. *Ann. Sci. Éc. Norm. Supér. (4)* **55** (2022), no. 5, 1289–1328

[162] L. Ma and M.-C. Hong, Curvature flow to the Nirenberg problem. *Arch. Math. (Basel)* **94** (2010), no. 3, 277–289

[163] A. Malchiodi, The scalar curvature problem on S^n: An approach via Morse theory. *Calc. Var. Partial Differential Equations* **14** (2002), no. 4, 429–445

[164] A. Malchiodi and M. Mayer, Prescribing Morse scalar curvatures: Subcritical blowing-up solutions. *J. Differential Equations* **268** (2020), no. 5, 2089–2124

[165] A. Malchiodi and M. Mayer, Prescribing Morse scalar curvatures: Blow-up analysis. *Int. Math. Res. Not. IMRN* (2021), no. 18, 14123–14203

[166] A. Malchiodi and M. Mayer, Prescribing Morse scalar curvatures: Pinching and Morse theory. *Comm. Pure Appl. Math.* **76** (2023), no. 2, 406–450

[167] A. Malchiodi and M. Struwe, Q-curvature flow on S^4. *J. Differential Geom.* **73** (2006), no. 1, 1–44

[168] F. C. Marques, Existence results for the Yamabe problem on manifolds with boundary. *Indiana Univ. Math. J.* **54** (2005), no. 6, 1599–1620

[169] W. S. Massey, *A basic course in algebraic topology*. Grad. Texts Math. 127, Springer, New York, 1991

[170] P. Mastrolia, M. Rigoli, and A. G. Setti, *Yamabe-type equations on complete, noncompact manifolds*. Prog. Math. 302, Birkhäuser/Springer Basel AG, Basel, 2012

[171] D. Maxwell, Solutions of the Einstein constraint equations with apparent horizon boundaries. *Comm. Math. Phys.* **253** (2005), no. 3, 561–583

[172] M. Mayer, Prescribing scalar curvatures: Non compactness versus critical points at infinity. *Geom. Flows* **4** (2019), no. 1, 51–82

[173] M. Mayer, Prescribing Morse scalar curvatures: Critical points at infinity. *Adv. Calc. Var.* **15** (2022), no. 2, 151–190

[174] M. Mayer, Prescribing Morse scalar curvatures: Incompatibility of non existence. 2021, arXiv:2106.09294

[175] M. Mayer and C. B. Ndiaye, Barycenter technique and the Riemann mapping problem of Cherrier–Escobar. *J. Differential Geom.* **107** (2017), no. 3, 519–560

[176] M. Mayer and C. Zhu, Prescribing scalar curvatures: On the negative Yamabe case. 2023, arXiv:2302.02435

[177] J. Milnor, *Lectures on the h-cobordism theorem*. Princeton University Press, Princeton, NJ, 1965

[178] J. Moser, On a nonlinear problem in differential geometry. In *Dynamical systems (Proc. Sympos., Univ. Bahia, Salvador, 1971)*, pp. 273–280, Academic Press, New York, 1973

[179] M. Musso and A. Pistoia, Tower of bubbles for almost critical problems in general domains. *J. Math. Pures Appl. (9)* **93** (2010), no. 1, 1–40

[180] C. B. Ndiaye, Y. Sire, and L. Sun, Uniformization theorems: Between Yamabe and Paneitz. *Pacific J. Math.* **314** (2021), no. 1, 115–159

[181] M. Obata, The conjectures of conformal transformations of Riemannian manifolds. *Bull. Amer. Math. Soc.* **77** (1971), 265–270

[182] B. Osgood, R. Phillips, and P. Sarnak, Compact isospectral sets of surfaces. *J. Funct. Anal.* **80** (1988), no. 1, 212–234

[183] B. Osgood, R. Phillips, and P. Sarnak, Extremals of determinants of Laplacians. *J. Funct. Anal.* **80** (1988), no. 1, 148–211

[184] R. S. Palais, The principle of symmetric criticality. *Comm. Math. Phys.* **69** (1979), no. 1, 19–30

[185] S. M. Paneitz, A quartic conformally covariant differential operator for arbitrary pseudo-Riemannian manifolds (summary). *SIGMA Symmetry Integrability Geom. Methods Appl.* **4** (2008), article no. 036

[186] A. Pistoia and C. Román, Large conformal metrics with prescribed scalar curvature. *J. Differential Equations* **263** (2017), no. 9, 5902–5938

[187] J. Qing and D. Raske, On positive solutions to semilinear conformally invariant equations on locally conformally flat manifolds. *Int. Math. Res. Not.* (2006), article no. 94172

[188] A. Ratto, M. Rigoli, and L. Véron, Conformal immersions of complete Riemannian manifolds and extensions of the Schwarz lemma. *Duke Math. J.* **74** (1994), no. 1, 223–236

[189] A. Rauzy, Courbures scalaires des variétés d'invariant conforme négatif. *Trans. Amer. Math. Soc.* **347** (1995), no. 12, 4729–4745

[190] O. Rey, The role of the Green's function in a nonlinear elliptic equation involving the critical Sobolev exponent. *J. Funct. Anal.* **89** (1990), no. 1, 1–52

[191] D. Ruiz, Conformal metrics of the disk with prescribed Gaussian and geodesic curvatures. 2021, arXiv:2108.12815

[192] J. Sacks and K. Uhlenbeck, The existence of minimal immersions of 2-spheres. *Ann. of Math. (2)* **113** (1981), no. 1, 1–24

[193] R. Schoen, Conformal deformation of a Riemannian metric to constant scalar curvature. *J. Differential Geom.* **20** (1984), no. 2, 479–495

[194] R. Schoen, Notes by D. Pollack from a graduate course at Stanford in 1988. https://sites.math.washington.edu/~pollack/research/Pollack-notes-Schoen1988.pdf

[195] R. M. Schoen, Variational theory for the total scalar curvature functional for Riemannian metrics and related topics. In *Topics in calculus of variations (Montecatini Terme, 1987)*, pp. 120–154, Lect. Notes Math. 1365, Springer, Berlin, 1989

[196] R. M. Schoen, On the number of constant scalar curvature metrics in a conformal class. In *Differential geometry*, pp. 311–320, Pitman Monogr. Surveys Pure Appl. Math. 52, Longman Scientific & Technical Publisher, Harlow, 1991

[197] R. Schoen and S. T. Yau, Existence of incompressible minimal surfaces and the topology of three-dimensional manifolds with nonnegative scalar curvature. *Ann. of Math. (2)* **110** (1979), no. 1, 127–142

[198] R. Schoen and S. T. Yau, On the proof of the positive mass conjecture in general relativity. *Comm. Math. Phys.* **65** (1979), no. 1, 45–76

[199] R. Schoen and S. T. Yau, Proof of the positive mass theorem. II. *Comm. Math. Phys.* **79** (1981), no. 2, 231–260

[200] R. Schoen and S.-T. Yau, Conformally flat manifolds, Kleinian groups and scalar curvature. *Invent. Math.* **92** (1988), no. 1, 47–71

[201] R. Schoen and S.-T. Yau, Positive scalar curvature and minimal hypersurface singularities. In *Surveys in differential geometry 2019. Differential geometry, Calabi-Yau theory, and general relativity. Part 2*, pp. 441–480, Surv. Differ. Geom. 24, International Press, Boston, MA, 2022

[202] R. Schoen and D. Zhang, Prescribed scalar curvature on the n-sphere. *Calc. Var. Partial Differential Equations* **4** (1996), no. 1, 1–25

[203] M. Schwarz, *Morse homology*. Prog. Math. 111, Birkhäuser, Basel, 1993

[204] W.-M. Sheng, N. S. Trudinger, and X.-J. Wang, The Yamabe problem for higher order curvatures. *J. Differential Geom.* **77** (2007), no. 3, 515–553

[205] M. Spivak, *A comprehensive introduction to differential geometry. Vol. IV*. 2nd edn., Publish or Perish, Wilmington, DE, 1979

[206] S. Stolz, Simply connected manifolds of positive scalar curvature. *Ann. of Math. (2)* **136** (1992), no. 3, 511–540

[207] M. Struwe, A global compactness result for elliptic boundary value problems involving limiting nonlinearities. *Math. Z.* **187** (1984), no. 4, 511–517

[208] M. Struwe, A flow approach to Nirenberg's problem. *Duke Math. J.* **128** (2005), no. 1, 19–64

[209] M. Struwe, *Variational methods*. 4th edn., Ergeb. Math. Grenzgeb., 3. Folge 34, Springer, Berlin, 2008

[210] G. Talenti, Best constant in Sobolev inequality. *Ann. Mat. Pura Appl. (4)* **110** (1976), 353–372

[211] N. S. Trudinger, Remarks concerning the conformal deformation of Riemannian structures on compact manifolds. *Ann. Scuola Norm. Sup. Pisa Cl. Sci. (3)* **22** (1968), 265–274

[212] N. S. Trudinger and X.-J. Wang, The intermediate case of the Yamabe problem for higher order curvatures. *Int. Math. Res. Not. IMRN* (2010), no. 13, 2437–2458

[213] R. C. A. M. Van der Vorst, Best constant for the embedding of the space $H^2 \cap H_0^1(\Omega)$ into $L^{2N/(N-4)}(\Omega)$. *Differential Integral Equations* **6** (1993), no. 2, 259–276

[214] M. Vaugon, Transformation conforme de la courbure scalaire sur la sphère. *Ann. Inst. H. Poincaré Anal. Non Linéaire* **3** (1986), no. 1, 55–65

[215] J. A. Viaclovsky, Conformal geometry, contact geometry, and the calculus of variations. *Duke Math. J.* **101** (2000), no. 2, 283–316

[216] J. Wei and X. Xu, Prescribing Q-curvature problem on $\mathbf{S}^n$. *J. Funct. Anal.* **257** (2009), no. 7, 1995–2023

[217] E. Witten, A new proof of the positive energy theorem. Comm. Math. Phys. **80** (1981), no. 3, 381–402

[218] H. Yamabe, On a deformation of Riemannian structures on compact manifolds. *Osaka Math. J.* **12** (1960), 21–37

[219] L. Zhang, Refined asymptotic estimates for conformal scalar curvature equation via moving sphere method. *J. Funct. Anal.* **192** (2002), no. 2, 491–516

[220] W. P. Ziemer, *Weakly differentiable functions*. Grad. Texts Math. 120, Springer, New York, 1989

